It is hard to think of any significant aspect of our lives that is not influenced by what we have learned in the past. Of fundamental importance is our ability to learn the ways in which events are related to one another, called associative learning. This book provides a fresh look at associative learning theory and reviews extensively the advances made over the last twenty years. *The psychology of associative learning* begins by establishing that the human associative learning system is rational in the sense that it accurately represents event relationships. David Shanks goes on to consider the informational basis of learning, in terms of the memorisation of instances, and discusses at length the application of connectionist models to human learning. The book concludes with an evaluation of the role of rule induction in associative learning.

This will be essential reading for graduate students and final year undergraduates of psychology.

The psychology of associative learning

Problems in the Behavioural Sciences

The psychology of associative learning

David R. Shanks
University College London

CAMBRIDGE
UNIVERSITY PRESS

Published by the Press Syndicate of the University of Cambridge
The Pitt Building, Trumpington Street, Cambridge CB2 1RP
40 West 20th Street, New York, NY 10011-4211, USA
10 Stamford Road, Oakleigh, Melbourne 3166, Australia

First published 1995

Printed in Great Britain at the University Press, Cambridge

A catalogue record for this book is available from the British Library

Library of Congress cataloguing in publication data

Shanks, David R.
The psychology of associative learning / David R. Shanks.
  p.   cm. – (Problems in the behavioural sciences; 13)
Includes bibliographical references and index.
ISBN 0 521 44515 9. – ISBN 0 521 44976 6 (pbk.)
1. Learning, Psychology of.   2. Association of ideas.   3. Memory.
4. Connectionism.   I. Title.   II. Series.
BF318.S43   1995
153.1'526 – dc20   95-3429   CIP

ISBN 0 521 44515 9 hardback
ISBN 0 521 44976 6 paperback

Contents

traced back to information presented to the senses. More importantly, it denies that organisms arrive in the world with significant amounts of innate knowledge already in place. I do not know whether empiricism will turn out to be correct, but often it seems that it is rejected (for instance, in Chomskyan linguistics) simply because no-one has managed to work out how a certain concept or piece of knowledge (such as a rule of grammar) could have been learned. But that is surely a defeatist attitude. The extraordinary learning capabilities of connectionist systems, described in Chapter 4, should make us very cautious about abandoning an otherwise useful default assumption.

I should also point out that as far as I am concerned, this book is as much about memory as about learning. It is regrettable that these topics are often seen as independent, for surely they are opposite sides of the same coin? 'Memory' simply describes the states that intervene between learning and behaviour. While it is true that I do not discuss many of the issues (e.g. short-term memory, retrieval) that occupy the minds of memory researchers, I do believe that the theoretical project of learning theory is the same as that of the study of memory: namely, to understand the processes by which information about the world is acquired and deployed by the human cognitive system.

A number of colleagues have been kind enough to read drafts of the book and offer their suggestions for improving it. Among them are Nick Chater, Anthony Dickinson, Zoltan Dienes, and Steven Sloman. I would like to thank them warmly for their helpful comments while also absolving them of any responsibility for errors or misconceptions in the end-product. I am particularly grateful to Tony Dickinson for getting me interested in learning in the first place, for being such an excellent collaborator subsequently, and for devoting so many hours to discussing learning theory with me. Finally, I should like to express my appreciation to the many students at Cambridge and University College who served as guinea pigs during the development of my often incoherent views on learning and memory.

The figures present data either taken from published tables or estimated graphically from published figures.

Preface

The place of learning theory in psychology has fluctuated dramatically over the last few decades. At the height of the behaviourist era in the 1940s and 1950s, many people would have identified learning as the single most important topic of investigation in psychology. By the 1970s, though, the picture was very different. A large number of influential psychologists had come to regard learning theory as a sterile field that made little contact with the realities of human cognition, and instead their interest switched towards such topics as knowledge representation and inference. Lately, things have come full circle, and the expanding new field of connectionism has restored learning theory to centre stage.

My aim in this book has been to have a fresh look at learning theory. I believe that the changing fortunes of the field have occurred because psychologists have often come to the topic of learning with certain deep conceptions of the mind already in place. But I suggest that rather than seeing learning as a topic to be annexed by and interpreted in terms of whatever the current fashionable theory happens to be, it is far more profitable for traffic to flow in the opposite direction. Let us commence by asking some general questions about learning, and see what sort of mind we end up with.

Someone coming to the field of learning afresh is likely to be disconcerted by the apparent incompatibility of a number of theoretical approaches and even terminologies. For instance, what relation is there between implicit learning and contingency judgment? What relation is there between a conditioned stimulus and a prototype? My overriding aim in this book has been to try to present a unified account of learning. I try to avoid the use of confusing terminology, and argue that the various theoretical approaches that have been offered can readily be compared with one another. The key idea is that different theories often look at things from different perspectives, one focusing (say) on a high-level description of what a learner is doing, while another concentrates on the details of th' learning mechanism. If we start out with a clear idea of the different que' tions we want to ask, and the different levels at which these questions m' be pitched, many of the confusions are likely to dissolve.

Of course, no-one can completely avoid carrying theoretical bag' around, so I might as well come clean and confess that I believe empiri' is the basic starting point from which learning should be vi' Empiricism is the doctrine that all true knowledge of the world '

1   Introduction

It is hard to think of any significant aspect of our lives that is not influenced by what we have learned in the past. The world looks and sounds the way it does because as infants we learned to partition it up in certain meaningful ways: we see familiar faces rather than meaningless blobs of colour and hear words rather than noise. Similarly, we behave in the ways we do because we have learned from past experience that our various actions have certain specific consequences. Like many topics of psychological inquiry, the importance of learning can perhaps best be realised by considering what life is like for people who have learning difficulties. Consider the case of Greg, a patient described by Sacks (1992), who became profoundly amnesic as a result of a benign brain tumour that was removed in 1976. Although his memory for events from his early life was almost completely normal, Greg remembered virtually nothing that had happened to him from 1970 onwards and appeared quite unable to learn anything new. He continued to believe, for instance, that Lyndon Johnson was the American President. In 1991 he was taken to a rock concert given by a group that he had been a great fan of in the 1960s, and despite sitting through the concert in rapture and recalling many of the songs, by the next morning he had no memory of the concert. More distressingly, when told of his father's death he was immeasurably sad but forgot the news within a few minutes. He was unable to learn that his father was no longer alive, and relived his grief anew every time he was told the news.

Another of Greg's difficulties, more mundane but also more representative of the sorts of behaviours studied by psychologists, was his inability to remember lists of words for more than a few minutes. This difficulty had nothing to do with understanding the words, since Greg's linguistic knowledge, having been learned early in life, was preserved. Rather, it is attributable to an inability to learn new associations between a certain context – the time and place in which the list was read – and the words on the list. For a normal person, remembering a list of words heard in a certain context would be relatively straightforward, since recollection of the context would bring the words to mind simply by association. Greg's other problems, such as his inability to learn who is President, can also be interpreted in terms of a basic difficulty in learning associations: a normal person will have little difficulty learning a new association between a name and the label 'President'.

Although learning lists of unrelated words may not be of much value in

1

the real world, in general this sort of *associative* learning is at the heart of any organism's psychological capabilities, because it endows the organism with the ability to adapt its behaviour as a result of acquiring information about associations or contingencies that exist between events in its environment. The ability to search out rewards like food and avoid threats like predators can only be achieved by learning predictive relations between rewards and threats, on the one hand, and events that are reliable signals of them, on the other. And as many researchers have observed, this adaptive ability is a major feature of what we understand by terms like 'intelligence' and 'intelligent behaviour.' Indeed, as Howard (1993) has pointed out, one widespread definition of intelligence equates it with flexible learning: a person (or animal) who is able to learn efficiently and transfer its knowledge to new situations that it encounters is called intelligent.

This book reviews research conducted over the last 20 or so years on the psychology of human learning and focuses specifically on associative learning. In an associative learning situation, the environment (or the experimenter) arranges a contingent relationship between events, allowing the person to predict one from the presence of others. The predictive events will either be external signals which I shall term 'cues', or the subject's own actions. Predictive relationships can be of two sorts, causal or structural. The most obvious form of relationship is *causal*, where one event or set of events is followed after an interval of time by another. For instance, in my office there is (barring an electrical fault) a consistent causal relationship between pressing the light switch and the light coming on. In contrast, we may say that a relationship is *structural* when an organism learns to predict one feature or attribute of an object or event from the presence of other features that regularly co-occur with it. For example, as a result of exposure to the co-occurrence of the sight and sound of running water, an organism may benefit from being able to predict that the sound of water is a good index of the sight of it. The ability to classify objects is another example of structural prediction. When I classify a particular sound as a word I am assigning it to a category, the category consisting of all the possible tokens of that spoken word. But the relationship between sound and category is structural, not causal: being a member of the category is a feature or property of the sound.

The term 'associative learning' has traditionally been meant to provide a contrast with 'nonassociative learning', but in fact this contrast is probably of little significance. Typically, the term nonassociative learning has been used to describe phenomena such as habituation, priming, and perceptual learning where, in contrast to associative situations, no explicit contingencies between experimenter-defined stimuli or actions are programmed, but where learning can nevertheless be observed. Thus in perceptual learning, subjects are simply exposed to isolated stimuli (such as faces) and learn to discriminate them better than would otherwise be possible, without there being any overt contingency between these stimuli and other events. The problem with this definition is that there inevitably are structural contingencies amongst the *elements* of a stimulus, and so as researchers like McClelland and Rumelhart (1985) and McLaren, Kaye and Mackintosh (1989) have noted, so-called nonassociative learning may be grounded in associative learning of those contingencies. I shall not discuss in any detail tasks such as priming that are typically regarded as examples of nonassociative learning, but it is worth bearing in mind that the principles of associative learning may be perfectly applicable to nonassociative learning as well.

What exactly is meant by 'learning'? The definition of this apparently innocuous term has been a topic of passionate debate by psychologists. In their enthusiasm to rid the subject of mentalistic concepts, the behaviourists argued that learning must be observable and that therefore it should be equated with the emergence of new patterns of responding. When we say that a dog in a laboratory Pavlovian conditioning experiment has learned something about the relationship between a bell and food, what we mean is just that a new behaviour has been conditioned: the dog salivates to the bell, whereas previously it did not. On such a view, we should only use the term 'learning' if there is some observable change in behaviour, in which case the new behaviour *is* the learning.

However, there are at least two obvious problems with this definition. The first is that learning may occur without any concomitant change in behaviour: if a cue and an outcome such as shock are presented to subjects administered drugs that block muscular activity, conditioned responding may perfectly well occur to the conditioned stimulus when the paralytic drug has worn off (e.g. Solomon and Turner, 1962). Learning clearly occurs when the animals are paralysed, even though no behavioural changes take place at that time. The second problem is that in many cases it can be established that organisms do much more than simply acquire new types of behaviour. For instance, in a famous experiment, MacFarlane (1930) trained laboratory rats to run through a maze to obtain food, and found that when the maze was filled with water, the animals continued to obtain the food even though they now had to swim to reach it. Clearly, learning in this case does not merely involve the acquisition of a set of particular muscle activities conditioned to a set of stimuli, but instead involves acquiring knowledge of the spatial layout of the maze, with this knowledge capable of revealing itself in a variety of different ways.

For these reasons, a more cognitive view is that learning is an abstract term that describes a transition from one mental state to a second in which encoded information is in some way different. This transition may perfectly well take place without the development of any new behaviour, and furthermore may manifest itself in a variety of quite different behaviours. But although it avoids the problems associated with defining learning in terms of behaviour, this definition also has its shortcomings. For instance, how are we to distinguish between learning and forgetting? Forgetting, like learning, can be viewed as a change in encoded information, except that in this case information is lost rather than gained. Our definition would plainly need to be supplemented by a proviso that learning involves the gain of information, but it is likely to be very difficult to specify what we mean by 'information gain' without relying simply on behaviour. We might find ourselves reverting to a behavioural definition of learning, which is precisely what we were trying to avoid.

Another problem with the cognitive definition is that it fails to deal satisfactorily with examples of what we might call 'non-cognitive' learning. The cognitive definition refers to a transition from one *mental* state to another, and the reason for incorporating the restriction to mental states is to exclude examples like the following. Roediger (1993) reports that the average duration of labour for first-born babies is about 9.5 hours, while that for later born babies is about 6.6 hours. Clearly, for second and third children the amount of time the mother spends in labour is much less than for first children. It seems strange to say that the female reproductive system is capable of 'learning' and 'remembering', so we would like to exclude this sort of case, despite the fact that information has obviously been acquired by the body. The restriction to mental states excludes the labour case because the relevant changes take place in the body without any mental component. But then we seem to be committed to saying that all habits and skills (which we do want to include as examples of learning) must be mental, and this seems unduly restrictive. Is it not likely that some aspects of learning a skill like playing tennis are really bodily rather than mental? Borderline cases like this probably illustrate the futility of trying to define learning.

Since the experimental study of learning began in the late nineteenth century, when Hermann Ebbinghaus (1885) commenced his pioneering laboratory investigations of human learning and evolutionists such as George Romanes (1882) began to use controlled experiments to investigate animal intelligence, the study of animal and human learning has continued in parallel, but regrettably not always with as much cross-reference as one might wish. Although in this book I focus solely on human learning, I hope that nothing concluded here will offend a student of animal learning. While humans may have learning capabilities that are available to few (or no) other organisms, such as the ability to abstract general rules (Mackintosh, 1988), I would argue that the correspondences between human and animal learning mechanisms far outweigh their differences. Some psychologists believe that since the motivation and prior experience of laboratory animals can be carefully controlled, it is research with animals that tends to produce the major discoveries about learning, with the study of human learning merely following along behind.

However, one of the principal aims of this book is to show that genuine insights about learning have been made in the last decade or two of research on humans. Because amenable human subjects only require appropriate instructions in order to perform almost any task, no matter how bizarre, there are a number of things that can be investigated in humans that would be extremely difficult, if not impossible, to study in non-humans. Obvious examples include tasks requiring subjects to make similarity and probability judgements, from which a wealth of interesting findings have emerged. Further, data can be obtained from humans that are orders of magnitude more complex, and therefore theoretically challenging, than those obtainable from non-humans. The obvious example involves language acquisition, where even highly simplistic models need to be of great sophistication (e.g. Elman, 1990; Pinker, 1991). In addition to focusing solely on human learning, the discussion in this book is also restricted in that it will not extend to language learning. To cover language acquisition would of course require a book in itself, but I should mention that I believe that the ability to explain language learning is the touchstone of any theory of learning, and I would be surprised if language learning turns out to rely on mechanisms radically different from those discussed here.

Historical background

For much of the century following Ebbinghaus' (1885) pioneering studies of learning, research has been conducted either explicitly or implicitly within the associationist tradition. It is important to distinguish the term 'associative' as in 'associative learning' from the term 'associationism.' The former is a purely descriptive term referring to the type of learning that takes place – whatever its nature – when a relationship exists between certain events in the organism's environment. 'Associationism', in contrast, refers to a particular view of how that learning is effected: it is the view that in the final analysis, all knowledge is based on connections between ideas. Sensory systems provide an organism early in life with very simple perceptual experiences, which during development become associated as a result of their co-occurrence to yield more complex experiences. These associations are such that when one has been formed, it automatically carries the mind from one idea to another.

Associationism took a central place in the psychology of learning not only because of its simplicity but also, and more fundamentally, because it provided the bedrock for the empiricist analysis of the mind, and of science in general, which had become dominant by the nineteenth century. An obvious difficulty for the empiricist view that all knowledge is derived from experience is that many concepts or ideas, such as a biologist's concept of a gene, are infinitely more complex than the simple sensory experiences that, according to empiricism, provide the only foundation for knowledge. Associationism provides a potential solution in the hypothesis that primitive experiences can become associated to yield more complex ideas, and those ideas can then in turn become associated to yield even more complex ideas, until all of the complexity of the biologist's concept is accounted for.

The associationist perspective provided both an overall conception of learning and knowledge, and also in the hands of Ebbinghaus and his followers the obvious means of investigating learning. If simple association of ideas is the only process involved, then all that is needed is to set up some simple associative learning task in which as many superfluous features as possible are removed, and use it to investigate the basic laws of learning. The learning of lists of nonsense syllables and of 'paired-associates' provided the ideal solution: learning that the nonsense syllable *wux* was on a list seems to require nothing more than the formation of associations between the phonetic elements of the syllable and associations between those elements and the list context, while learning the arbitrary response 'reason' to the stimulus word 'window' requires the formation of an associative bond between two pre-existing but previously unconnected ideas. Thus it was thought that laboratory studies of nonsense syllable and paired-associate learning would be sufficient to uncover *all* of the universal laws of learning.

By the 1950s the associationist analysis of learning had reached almost total dominance, to the point where many textbooks on learning and memory took it for granted that associationism provided the only explanatory framework worth considering. In the hands of researchers like Benton Underwood (1957) and Leo Postman (1962), paired-associate learning acquired the status in studies of human learning that the Pavlovian conditioning procedure has acquired in animal research and, as is clear from reviews such as that of Deese and Hulse (1967), sophisticated discussions of whether learning occurred gradually or was all-or-none, whether forgetting was due to trace decay or interference, seemed to imply that genuine progress was being made.

However, by the early 1970s cognitive psychologists had begun to tire not only of such artificial tasks as paired-associate learning, but also of learning in general. Partly, no doubt, this was due to the apparent intractability of some of the key issues: investigators had argued themselves to a standstill over the continuous versus all-or-none debate, for instance (Restle, 1965). But more important was a long-term shift of interest towards knowledge representation. Two particular strands to this shift are worth considering.

The associationist view that complex concepts can be reduced to the association of elementary ideas had been resisted throughout the first half of the century by a small minority of researchers, including Gestaltists such as Koffka (1935) and Kohler (1947). Rather than consisting of the association of ideas, the Gestaltists emphasised the importance of organisation, and viewed learning as the construction of an organised whole in which the associated items are subcomponents. On this view, learning does not proceed via the strengthening of simple connections between ideas: rather, it involves the construction of new entities, holistic memory traces representing the ideas, their conjunction, and the current context. Retrieval, similarly, does not involve the activation of one idea via the flow of energy along a connection, but rather the reactivation of an entire memory trace.

Moreover, it was not only the Gestaltists who took this view of learning. In a paper published in 1893, the philosopher James Ward asked the apparently simple question of why repetition improves memory, and challenged the typical associationist view that repetition leads to the gradual strengthening of a mental bond or connection. Instead, Ward proposed that each repetition lays down a quite separate memory trace, and that memory improves because more traces exist to be accessed. The largely-forgotten memory researcher Richard Semon also advocated such a multiple-trace view (see Schacter, Eich and Tulving, 1978), and combined it with sophisticated ideas about how retrieval occurs (he coined the term 'ecphory' that refers to the reactivation of a complete memory trace on presentation of a cue that matches part of it). The multiple-trace view has, of course, been continued in recent years both in Endel Tulving's (1983) work on memory retrieval and in Hintzman's (1976) use of frequency judgments to try to discriminate strength and multiple-trace views of repetition effects.

The holistic view of learning and representation gained support from a large number of animal discrimination–learning experiments conducted during the 1950s. Suppose an animal is shown two red stimuli on some trials and is rewarded for choosing the right-hand one, while on other trials, a pair of green stimuli is presented and reward is given for choosing the left-hand stimulus. Simple though this discrimination may be, it cannot be solved on the basis of associations between the simple co-occurring elements of the task (red, green, left, right, reward, non-reward). This is because each element should become equally associated with each other element: red and green, for instance, are equally associated with left and right and reward and non-reward. The fact that humans as well as laboratory rats can learn these discriminations (e.g. Bitterman, Tyler and Elam, 1955) argues that the simplest sort of associationist analysis is insufficient, although in Chapter 4 we will see that this sort of discrimination, which is called a *nonlinear* classification, can be dealt with by more modern associationist theories.

In contrast to the difficulty posed for associationism, Gestalt views of learning are sufficiently flexible to be untroubled by this sort of discrimination learning: the organism is assumed to memorise the entire set of elements occurring on a given trial, such as {red, right, reward}, with the elements merely being parts of a larger memory trace. The organism may now solve the discrimination when shown a red stimulus by recalling that *right* is the choice that has been rewarded {red, ?, reward}. This sort of analysis led Medin (1975) to formulate an explicit model of configurational learning in the Gestaltist tradition called the context model, and we shall see in Chapter

3 that this theory has had some striking successes in describing human learning data. The model adopts a radically different unit of analysis from that of traditional associationist accounts: instead of elements becoming associated with outcomes, it is memorised configurations or 'instances' that underlie learning. The overall degree of similarity between a test item and the ensemble of stored instances determines the response that the item evokes.

The second and perhaps more powerful reason for an increasing interest in knowledge representation arose from the advent of the computer as a new model of the mind which made associationism seem totally inadequate. With the development of knowledge-representation programming languages like Lisp, investigators such as Newell, Shaw and Simon (1958) quickly began to develop computer models of highly complex human abilities such as solving logic problems and playing chess. Not only must paired-associate learning have looked decidedly trivial in comparison, but also the explanatory apparatus of these new computational models was far richer than associationism allowed: languages like Lisp represent knowledge symbolically, which meant that inference was readily possible, and the success of these models clearly suggested that their symbolic data-structures corresponded to those that were actually used by the human mind. The idea that complex knowledge, as argued by Quillian (1968), consists of concepts connected in propositional networks by semantically-interpreted relations appears to be quite at variance with associationism.

The development of these richer views of knowledge representation had a concomitant influence on ideas about learning. If knowledge is represented propositionally, then learning must involve the construction of propositional structures via a set of pre-existing general symbol-manipulating procedures. Accordingly, as computer models evolved, it became increasingly popular to view learning as a form of hypothesis-testing or rule-induction, and detailed studies of rule-induction were carried out, most famously by Bruner, Goodnow and Austin (1956) and Hunt, Marin and Stone (1966). As we will see in Chapter 5, such an approach has continued to this day and offers some important insights into learning.

Not everyone was persuaded by the rule-induction view of learning, and dissatisfaction was greatest amongst researchers interested not so much in learning as in conceptual representation. In a seminal article, Eleanor Rosch (1973) pointed out that knowledge of everyday objects such as chairs and birds is unlikely to be based on inductively-learned rules, since such rules would have to specify certain necessary and sufficient features for an object to be a member of the category. Yet surely, she argued, no such features exist: what could the necessary and sufficient perceptual features possibly be that define the category *bird*? Rather than sharing a set of common defining features, each member of a category can be thought of as a set of features, with large degrees of overlap between the features of different members of the category but with none of the features being necessary or sufficient. On this view, categories may show 'graded structure', with some members of the category possessing more of its characteristic features than others and hence being more typical. As Barsalou (1990) has shown, such graded structure is a property of almost all categories, a fact that encourages the view that categories are represented by mental prototypes which correspond to objects possessing all of the characteristic features of that category.

Prototype theories therefore suggest that the learning process involves abstracting the category prototype from the experienced exemplars. A novel item is classified according to its similarity to the prototype stimulus. We will consider the prototype approach in Chapter 3, but here it is worth briefly mentioning one historical development that played a major role in the construction and testing of the instance and prototype theories discussed in Chapter 3. Objects in the world vary on a large number of independent dimensions such as colour, size, height, and so on, and we can therefore represent each object as a point in a multi-dimensional physical space. Each object also corresponds to a point in a corresponding mental space, where the dimensions of the space are those that the perceptual system uses to represent stimuli. However, the physical and psychological spaces may not be at all similar. For example, colour is one of the most salient aspects of mental representation, but has no exact physical correlate (see Hardin, 1990) – it is a psychological property. In the late 1950s, researchers began to develop the tools needed to analyse psychological spaces. By obtaining *proximity* measures such as similarity ratings from all pairwise combinations of a set of stimuli, it is possible to recover the locations and organisation of the stimuli in psychological space using statistical procedures such as multi-dimensional scaling and cluster analysis (see Shepard, 1980). From the point of view of learning, these developments have had immense importance. As a consequence of a learning episode, some representation of the training items will be formed, perhaps a prototype. Responding to a test item will be a function of its similarity to this representation. But how do we know how similar it is? Techniques such as multi-dimensional scaling provide an answer and therefore allow us to predict with great accuracy how test items will be treated.

Thus research up to the mid-1970s had set the scene for a variety of alternatives to the traditional associationist theories. Healthy interest was being paid to prototype and instance theories and to rule induction processes. The subsequent development of each of these approaches will be the focus of Chapters 3 and 5, while Chapter 4 will discuss the various ways in which associationism has evolved over the last two decades.

Three questions about learning

In this brief historical review, I have discussed alternative theories of learning as if they are incompatible with one another and as if evidence in favour of one theory must necessarily weaken the support of the others. There is no doubt that many researchers see the various theories in this way. Hintzman (1976), for example, was quite adamant that frequency judgment data supported the multiple-trace approach and disproved theories based on associative strength, while more recently Waldmann and Holyoak (1992) claimed that some data of theirs 'clearly refute connectionist learning theories that subscribe to an associationistic representation of events as cues and responses' (p. 233). But it is also possible to see these theories as differing in their level of analysis and thus as not necessarily incompatible. Perhaps processes which at one level of analysis are well-characterised as being associationistic can also be described, at another level, as involving prototype abstraction or instance memorisation?

Ever since Marr (1982) published his highly-influential analysis of levels of explanation, psychologists have had to consider quite carefully how their research questions should be framed. Marr distinguished very clearly between the questions of what the system is computing and how it is doing it. Such questions need to be answered in quite different ways; the 'what?' question cannot be answered, for instance, by citing some complex brain mechanism which is more appropriate for the 'how?' question. With respect to learning, the highest level requires us to consider what it is in the environment that the associative learning system is sensitive to, while lower levels concentrate on the internal characteristics of the learning mechanism itself.

Following Anderson's (1990) extensive discussion of the different sorts of questions that can be asked of the cognitive system, I shall adopt the view that theories of learning have to address the following three fundamental questions. The first asks whether associative learning is *normative* (and hence rational). An associative relationship consists of a temporal distribution of events, and normative theories tell us whether or not an objective relationship exists in a given situation. Such theories can therefore be regarded as providing independent measures of association. Although the framing of normative theories is perhaps more the business of philosophers and statisticians than of psychologists, we will see that consideration of such theories is highly relevant to an understanding of learning. If it turns out that people perform well in comparison with the norms provided by a rational analysis, then it is reasonable to conclude that some mental algorithm exists for computing the norms in question. Chapter 2 considers the view, recently developed in detail by Cheng and Holyoak (1994), that the appropriate normative theory of associative learning is *contingency* theory. This theory provides a means of determining for any given situation what the objective relationship is between a pair of events; on this theory, people behave normatively or rationally if they believe events to be related only when contingency theory specifies that they indeed are. By considering evidence that has accumulated during the last 20 years, we will ask whether the human associative learning system behaves in ways that would be judged 'rational' given the prescriptions of contingency theory.

The first question establishes what the system is doing. Once this question is dealt with, we can ask the second question. How, in broad informational terms, is it doing it? What forms of representation are actually learned that allow the system to behave as it does? I shall call this the *representation* question. For instance, learning may involve the memorisation of relatively unanalysed experiences or the construction and testing of complex hypotheses. For each such theory, a different representational assumption is made concerning the sort of data-structure that is learned. In Chapter 3 we will consider two popular answers to this question that were briefly discussed above. The first proposes that learning is mediated by mental prototypes that are abstracted from a set of learning episodes, and the second argues that separate memory traces of each training item are stored. We will see that the memorisation of multiple events provides a very good explanation at the informational level of what the human learning system is doing.

The final question concerns the most basic level of analysis that psychology is concerned with. Given an adequate answer to the second question, we then want to know what the precise mechanism is that carries out these informational processes. For example, how does the system memorise instances? In Chapter 4 we will see that contemporary associationist models of learning provide detailed answers to this final *mechanism* question. Such models are mechanistic in the sense that they react moment-by-moment to the stimulus environment and learn associative relationships on-line, and they achieve this via processes such as activation and inhibition that are known to occur in the brain. Moreover, they show why it is that the system looks, from the informational perspective, as if it is encoding multiple memory traces and they also show how rational behaviour can emerge. Thus the approach adopted here is to ask a series of questions which progressively reveal more and more detail about the workings of the system. The first task is to characterise what the system is doing, the next is to describe how in broad informational or computational terms it is doing it, and the third is to explain the exact mechanism that achieves this computation.

Two examples may help to clarify matters. Let us suppose that a previously-untrained subject in a laboratory learning experiment has seen a series of electrocardiogram traces depicting normal and abnormal heart functioning and has attempted to learn what characterises the abnormal traces. Then, he or she is shown a new trace and asked to decide whether it is normal or abnormal. We wish to understand why the subject makes the decision he or she does. The normative question suggests that we should begin by asking what the subject should rationally do in this situation. For this, we need a normative theory of classification. The pattern classification literature has established (e.g. Duda and Hart, 1973) that correct classifications are approximately maximised when an object is placed into the category containing the training item most similar to the test item. Thus, we find the nearest neighbour to the test item amongst the training stimuli, and place the test item in the nearest neighbour's category (in this case, the categories are 'normal' and 'abnormal').

Having established what is rational for our subject to do, we see whether he or she behaves in accordance with this rational prescription. If the answer is 'yes' – and previous analyses (Anderson, 1990) suggest that it will be if we have thought carefully enough about our rational theory – we then go on to the second question and ask what information the subject needs to have encoded during the study phase in order to behave in this normative manner. Clearly, in order to perform the nearest-neighbour analysis, the subject must have memorised each training item together with its category. Finally, we can ask how this memorisation process can be implemented in a learning mechanism. We may, for instance, be able to construct a connectionist system that implicitly performs the appropriate memorisation process. At that point we would be approaching a complete psychological explanation of the subject's behaviour.

The second example concerns forgetting. Suppose we are interested in why forgetting occurs at the rate and in the precise manner that it does. Again, the initial question to ask is 'what is it rational to do?' At first sight, it might seem that any degree of forgetting over time is irrational, but if the memory system is limited in capacity, some loss of information over time will be necessary to allow new information to be stored. It turns out (Anderson and Schooler, 1991) that the normative thing to do is to forget as a decreasing power function of time, since that is the rate at which information in the environment becomes redundant. Anderson and Schooler demonstrated this, at least for one situation, by examining an electronic version of the *New York Times*. They found that if a certain name such as 'Qaddafi' appeared in a headline in the newspaper on a given day, then the likelihood that it would appear on a later day was a decreasing power function of elapsed time. Having established that this is what the system should do, we can then ask whether humans or other organisms do indeed forget at this rate (the answer appears to be 'yes'; see Wixted and Ebbesen, 1991). We can then proceed by considering what sort of informational processes (such as retroactive interference) would yield this behaviour, and finally what sort of processing mechanism could carry out these informational processes.

In sum, a complete psychological analysis of learning requires consideration of at least three separate questions, and I have labelled these the normative, representation, and mechanism questions.

Dependent measures

For much of its history the experimental study of learning adopted an extremely limited number of procedures, amongst which paired-associate learning has had pride of place. Thankfully, in the time that has elapsed since the heyday of the associationists in the 1950s, a rather richer set of techniques has been examined. As a very brief illustration, laboratory investigations have made serious attempts to develop procedures that are representative of real-world learning tasks such as how medical practitioners acquire knowledge of diseases from photographic illustrations (Brooks, Norman and Allen, 1991), how people learn different artists' styles (e.g. Hartley and Homa, 1981) and syntactic relations (Morgan, Meier and Newport, 1989), how they learn to recognise faces (Nosofsky, 1992), add and subtract numbers (Young and O'Shea, 1981), and many others. In addition, countless investigations have attempted to study learning in natural settings.

In addition to this variety of procedures, laboratory associative learning tasks have utilised a range of dependent measures which I shall briefly introduce. The critical events are either external stimuli (called cues) or the subject's own actions, and depending on the relationship in effect, learning may be manifest in different ways. Perhaps the simplest emerges in changes in rate of responding in an action–outcome learning task. When the experimenter arranges a relationship or 'contingency' between an action and some valued outcome, such as earning money, a subject is likely to increase his or her rate of responding in the same way that animals in operant conditioning experiments do. In such experiments it is also straightforward to ask subjects to make numerical judgments concerning the action–outcome relationship, usually on a rating scale from –100 to +100, where –100 indicates that the action prevents the outcome from occurring and +100 means that it causes the outcome. Rating scales of this sort have been used in very many of the experiments that will be considered.

In cue–outcome learning tasks, a rather wider range of dependent measures is available. Once again it is possible to ask subjects to judge a cue–outcome relationship on a rating scale. A second possibility, applicable if the outcome is a motivationally-significant event, is to measure the degree of Pavlovian conditioned responding. For instance, while being exposed to pairings of a tone and a small electric shock, changes in galvanic skin conductance on the palms of the hands can be used as an index of learning. A third possibility, applicable to category learning experiments, is to examine choice responses: the subject in the study phase of the experiment may be shown some paintings, say, by unfamiliar artists A and B. After learning to correctly classify the study items, the subject may then be shown new paintings by A and B and asked to classify these 'transfer' items. Less commonly used have been measures of response time, but again these are perfectly valid indices of learning.

Implicit learning

Given this variety of measures, it is important to consider whether conclusions drawn on the basis of one measure also extend to other measures. In this regard, a variety of studies of human *implicit learning* have advanced the controversial conclusion that dissociations can be obtained, specifically between performance measures such as rate of responding on the one hand and consciously-based judgments on the other. The most obvious example of the latter would be a verbal judgment about the extent of an associative relationship, where such a judgment – unless it is a pure guess – is likely to be grounded in a conscious belief about the relationship. But this sort of explicit measure is not restricted to verbal reports, as we will see below: other forms of response can also be interpreted as being based on conscious knowledge.

Compelling examples of dissociations between implicit and explicit response measures come from amnesic patients such a Greg who, while having little conscious awareness of even the recent past, can nonetheless learn new pieces of music, for example, and retain them over long periods of time (Sacks, 1992). Such dissociations, if they are valid, suggest that instead of there being a single knowledge source which can be examined by any of a variety of tests, there are multiple sources some of which can only influence certain response measures. If this is the case, then it would seem that multiple learning systems must exist in order to feed the different knowledge systems with input. Learning is said to be 'explicit' or 'declarative' when it is accessible to consciousness, and 'implicit' or 'procedural' when it is not.

As an illustration of an apparent dissociation in normal subjects between these dependent measures, consider an experiment by Willingham, Greeley and Bardone (1993). On each trial, an asterisk appeared at one of four locations on a computer screen and subjects had to press as fast as possible the response key appropriate for that location. After pressing the correct response key, the next stimulus was presented. Subjects were given instructions appropriate for a typical choice RT task, but in fact for some subjects there was a sequence underlying the selection of the stimulus on each trial. For these subjects, a 16-trial stimulus sequence containing four occurrences of each stimulus was repeated many times over, while for control subjects the situation was identical except that the order of stimuli within each of the 16-trial sets was randomised.

Willingham *et al.* observed that reaction times dropped across 420 training trials, and that this speed-up was greater for the sequence than for the control subjects. Thus a performance measure – in this case reaction time – indicated that at least some aspects of the sequence had been learned by subjects exposed to the structured sequence. However, some of the subjects who showed this RT speed-up were not only unable to verbally report any of the predictive relationships in the stimulus sequence, but were also unaware that there had even been a sequence at all. Thus we appear to have evidence in these results of a dissociation between different dependent measures used to index learning: reaction times indicate sequence learning, whereas verbal judgments do not. Since a verbal judgment reports a conscious state of knowledge, many authors have been keen to interpret these dissociations as evidence for unconscious ('implicit') learning. Specifically, it has been claimed that measures indicative of conscious knowledge dissociate from those that do not require conscious knowledge (performance measures) and which instead may reveal unconscious sources of knowledge.

Another example, perhaps better-known, comes from studies of artificial grammar learning. An artificial grammar is a set of construction rules – vastly simplified compared to the grammar of a natural language – which is able to generate a set of strings. In most applications, such artificial grammars are used to generate strings of letters. Figure 1.1 shows such a grammar. To generate a string, the grammar is entered at the left-most node and a choice is arbitrarily made between the two available letters (M and V). If V is selected, the node at the end of the V path is reached and the options for the second letter are the paths out of that node, both of which add an X to the string. By moving though the network, letters are chosen until the right-most node is reached, at which point a complete grammatical sequence such as VXVRXR is generated.

Subjects are required to try to learn the structural rules of the grammar. In a typical experiment, such as that originally conducted by Reber (1967), they are presented with a list of grammatical strings in the first phase of the experiment and are told simply to learn the strings for a later memory test. Prior to that test, subjects are informed that the strings were formed according to a set of rules, and that they must now examine some new strings and decide which ones conform to the rules and which do not. Thus for each new test string, they have to make a grammaticality decision.

In order to perform above chance on such a test, it is obvious that subjects must learn something in the study phase concerning the structure of the stimuli. In fact, as we shall see in the forthcoming chapters, it is controversial as to what exactly subjects do learn. Nevertheless, the typical result is that subjects perform well above chance, usually being able to make about 65% correct decisions in the grammaticality test. But the interesting issue for our present purposes is that some of the knowledge used to decide whether an item is grammatical or not may be implicit and unavailable for conscious report by the subject. For instance, Mathews *et al.* (1989) presented subjects with study strings to memorise, and then in the test phase asked them to describe in exhaustive detail their reasons for deciding that a given test string was grammatical or nongrammatical. The verbal reports of these 'experimental' subjects were then given to 'yoked' subjects who had not been exposed to the study strings but who were required to judge the grammaticality of each test string by following and utilising the rules given to them by the experimental subjects. Mathews *et al.* found that although subjects in the experimental group had been exhorted to report all the structural rules they were aware of, the performance of the yoked subjects who were using those rules was considerably poorer than that of the experimental subjects themselves. Thus it seems that experimental subjects had uncon-

(a)   (b)

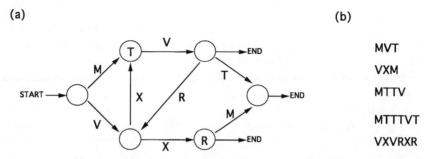

MVT

VXM

MTTV

MTTTVT

VXVRXR

Figure 1.1. (a) An artificial grammar for generating letter strings. To generate a string, the grammar is entered at the left-hand side and links are traversed until the grammar is exited on the right-hand side. Each link yields a new letter which is added to the string. If a node contains a letter, then the letter can be omitted or added repeatedly at that point in the string. (b) Some strings generated by the grammar.

scious access to certain bits of knowledge concerning the grammatical rules, but were unable to pass this knowledge on to the yoked subjects.

Dienes, Broadbent and Berry (1991) reached a similar conclusion using a rather more direct method. After presenting subjects with study strings to memorise, they then examined whether each subject's verbally reported rules were sufficient to account for his or her grammaticality decisions. The result was that the correlation between actual performance and performance predicted on the basis of reported rules was extremely low. Again, subjects appear to have unconscious access to knowledge of the grammar.

Despite this evidence for a distinction between conscious and unconscious learning, it is unclear at present whether such results demand an explanation in terms of distinct learning systems, and Mark St. John and I (Shanks and St. John, 1994) have argued that they may not. Three points need to be emphasised. First, there have been numerous examples of clear *associations* rather than dissociations between performance and report measures, which suggests that special conditions may be required to obtain dissociations. As a simple illustration, Shanks and Dickinson (1991) asked subjects to perform a simple instrumental learning task in which pressing a key on a computer keyboard was related, via a schedule of reinforcement, to a triangle flashing on the screen. Subjects were exposed to a reinforcement contingency in which they scored points whenever the triangle flashed, but lost points for each response, so that they were encouraged to adapt their response rate to the reinforcement schedule. Learning was demonstrated by changes in subjects' rates of responding. As a measure of awareness, subjects were asked to report on a scale from 0 to 100 what they thought the relationship was between the response and the reinforcer. We found that response rate and judgments were equally sensitive to two manipulations: first, as the degree of temporal contiguity between the action and outcome was reduced, both judgments and response rate declined, and secondly, a reduction of the degree of contingency between the action and outcome had comparable effect on judgments and response rate. These manipulations will be discussed at greater length in Chapter 2, but for present purposes all we need to observe is that the two types of response are often associated rather than dissociated.

Another example comes from a recent experiment by Lovibond (1992). He was interested in whether it is possible to dissociate conditioned responses to a conditioned stimulus (CS) that predicts an unconditioned stimulus (US), from conscious knowledge of the CS–US relationship. Lovibond presented subjects with two stimuli (slides depicting flowers or mushrooms), one of which (the CS+) was paired with shock while the other (CS–) was non-reinforced. Awareness of the relationship between the stimuli and shock was measured in two ways. First, during the learning phase subjects continually adjusted a pointer to indicate their moment-by-moment expectation of shock, and secondly, at the end of the experiment they were given a structured interview designed to assess their awareness.

At the end of the study Lovibond partitioned the subjects into those who apparently were aware and those who were unaware of the CS–US relationships. The left panel of Figure 1.2 shows that indeed some subjects (labelled 'unaware') gave no indication that they associated A with shock to a greater extent than B. Critically, the right-hand panel shows that these unaware subjects also gave no hint of stronger conditioned responding to A than to B. For subjects who were aware of the conditioning contingencies, galvanic skin responses (GSRs) were stronger to A than to B. Thus on these results we would have to conclude that learning about a CS–shock relationship does not occur in the absence of awareness of that relationship, and that this particular pair of dependent measures is hard to dissociate. It is also worth noting that when subjects are made necessarily unaware of associative relationships by the administration of general anaesthetic, associative learning as measured by performance indices also seems to be abolished. Thus Ghoneim, Block and Fowles (1992) were unable to obtain Pavlovian conditioning to an auditory stimulus in anaesthetised subjects using a procedure that elicited clear conditioning in awake subjects.

Of course, just because many situations fail to reveal dissociations between dependent measures does not mean that those situations, such as the serial RT task of Willingham *et al.*, where dissociations are obtained are any less genuine. However, there has been an extended debate about whether apparent dissociations reflect a psychologically-significant distinction between different knowledge sources, or merely reflect different degrees of sensitivity in different tests of knowledge, and this represents the second reason to question whether distinct learning systems exist. It remains a possibility that there is only a single source of knowledge, with different tests being able to detect different proportions of this knowledge. When an apparent dissociation between two tests emerges, perhaps it is because one of the tests has not detected information as exhaustively as the other.

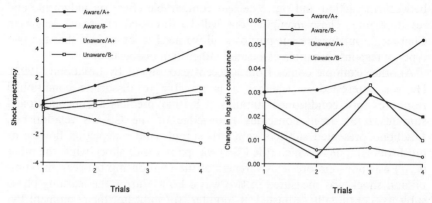

Figure 1.2. The relationship between contingency awareness and Pavlovian conditioning. Subjects were given four presentations each of stimuli A and B, with A being followed by shock and B being unreinforced, and shock expectancy and skin conductance were measured on each trial with each stimulus. The subjects were divided into two groups according to whether they were aware or not of the stimulus contingencies. The left-hand panel shows shock expectancy ratings for stimuli A and B in the Aware and Unaware groups, and the right-hand panel shows the change in skin conductance (the conditioned response) on each trial. Unaware subjects showed no evidence of differential conditioned responding while aware subjects did. (After Lovibond, 1992.)

As an illustration of the extent to which sensitivity may account for these dissociations, consider a further piece of data from Dienes *et al.*'s (1991) artificial grammar learning experiment described above. Recall that subjects were presented in the study phase with a set of letter strings generated from a grammar, and were then required in the test stage to discriminate between new grammatical and non-grammatical strings. Although Dienes *et al.* found that subjects' reportable knowledge was inadequate to account for their grammaticality performance, another test which Dienes *et al.* also interpreted as measuring conscious knowledge *was* adequate. In this additional test, subjects were shown incomplete letter sequences such as VXV... and asked to judge which letter continuations (...R? ...X?) were acceptable. The results from this continuation test correlated much better with grammaticality performance than did verbal reports; indeed, using a signal detection analysis, Dienes *et al.* found that the continuation test was at least as sensitive to knowledge of the composition rules as was the grammaticality test itself. Thus it is difficult at present to dismiss the claim that when dissociations between performance and conscious report measures of learning occur, this is because tests with differing degrees of sensitivity are being compared.

The third problem with claims for the existence of implicit learning concerns what Dunn and Kirsner (1989) have called the *transparency* assumption. So far, I have been assuming that any given test is either an implicit or an explicit one. For instance, in the Dienes *et al.* study the grammaticality test has been viewed as a test of implicit knowledge and the letter continuation test as a measure of explicit knowledge, and the question we have been considering is whether these tests yield similar or different results. But in order to conclude from a dissociation of two dependent measures that different mechanisms underlie performance on the two tasks, we must assume that performance of each task requires only a single psychological process. Yet this view – that a test is transparent with respect to a given psychological process – seems most unlikely to be true.

Instead, a given test is likely to engage a large number of diverse processes, some to a greater extent, some lesser. Granting for the moment that knowledge can be represented implicitly, it is likely that both the grammaticality and continuation tests in Dienes *et al.*'s study draw on both conscious and unconscious knowledge rather than depending exclusively on one form of knowledge or the other. On this conception, if there really are distinct conscious and unconscious learning processes, then it is extremely unlikely that transparent tests of them will exist. Instead, the existence of these processes will require a rather subtler form of experimental demonstration than merely attempting to find dissociations between two tasks. Some progress in this direction has been made (Jacoby, 1991), but it is clear that even if separate implicit and explicit learning processes exist, untangling them is going to prove very difficult.

Summary

The study of associative learning has been dominated during this century by associationist accounts which assume that the elements of a stimulus come to be related with one another and with the outcome as a result of their co-occurrence. Alternative approaches, including configurational and prototype theories, have also been advocated but these are better seen not as competitors to associationism but as answers to a different question. It is important to distinguish between the representational issue concerned with how, in informational terms, the learning system operates, and the implementational or mechanism question of how those computations are carried out. Prior to consideration of either of these questions, it is necessary to ask what the system is doing and whether it is normative or not.

This chapter has also briefly introduced some of the main ways in which associative learning is studied in the laboratory and has considered whether dissociations may be obtained between different response measures. Specifically, proponents of the existence of a separate implicit learning mechanism have argued that response measures which do not require conscious knowledge of an associative relationship may yield quite different results from measures which do. Although the data relevant to this issue are extremely interesting, I have suggested that as yet the existence of separate implicit and explicit learning systems has not been proved.

2 The rational analysis of learning

If we take the commonsense view that the human associative learning system has evolved for adaptive purposes, then it is immediately clear that the major benefits learning affords an organism are the ability to make predictions about events in the environment and the ability to control them. If it is possible to predict that a certain event signals either impending danger such as a predator, or imminent reward such as access to food, then appropriate action can be taken to avoid the danger or extract maximum benefit from the reward. It has been common, particularly in discussions of animal conditioning, to interpret learning from the perspective of the benefit it brings the organism.

Of course, this is not to say that learning is always beneficial, and the high incidence of maladaptive behaviours, such as phobias in humans that can be traced to prior learning episodes, attests to this fact. Nevertheless, it seems plausible that such learned behaviours emerge from a system that fundamentally exists to exploit and benefit from regularities that exist in the world, whether they be signal–outcome or action–outcome regularities. For instance, when Pavlov's dogs learned to salivate to a bell that signalled food, it is likely that they benefited from the increased digestibility and hence nutritional benefit of the food. When a child learns that saying 'juice' reliably produces a rewarding drink, it has acquired the ability to control a small but important aspect of its environment.

In this chapter we will try to establish the degree to which human learning is appropriately adapted to the environment. To the extent that people only learn associative relationships where they indeed exist, and do not believe events to be related when they are not, we can say that the system is well adapted. But how exactly are we to know whether a pair or set of events are objectively related? Clearly, we require some procedure or norm for specifying when events are truly related. In short, we need a normative theory which gives us a yardstick against which to compare human behaviour. The statistical concept called 'contingency' provides just such a yardstick, and the best-developed current theory of the objective relatedness of events relies on this notion of contingency. In this chapter we will examine contingency theory in some detail, particularly as it has recently been described by Cheng and Holyoak (1995), and will ask whether humans behave in associative learning tasks in the way prescribed by this normative model.

Information in the environment

The world provides a number of informational clues concerning associative relationships. Laboratory investigations using humans and other organisms have identified a number of such factors. In the present section we will look at what is perhaps the most fundamental of these, namely the degree of *correlation* or *contingency* over time between events. From a statistical perspective, associative learning simply requires the calculation of the degree to which a pair of events covary. In a typical task in which the subject attempts to work out whether two events are related, the two events can be treated as variables and the relationship between them can be described by some metric of statistical covariation. When the events are cues or actions and outcomes, then from a statistical point of view the variables are binary valued, that is, present or absent. The more general case is where the events are continuous variables which can vary in magnitude. For continuous variables, the term *correlation* is used to describe relatedness, while for discrete variables, the term *contingency* is used. It is now well-known that variations in correlation or contingency affect associative learning, and this is such a fundamental property of learning that we will devote a good deal of attention in forthcoming chapters to attempting to understand its basis.

For continuous variables, the appropriate measure of covariation is the correlation coefficient r. However, we will be more interested in binary-valued events, and Figure 2.1 shows a convenient way of representing such events that allows statistical contingency metrics to be straightforwardly calculated. Suppose we have some cue (or action) that is present or absent on each trial and is or is not accompanied by the outcome. The cells in the figure denote each of the possible combinations of events: the cue and the outcome (cell a), the cue but not the outcome (cell b), the outcome but not the cue (cell c), and the absence of both cue and outcome (cell d). A typical associative learning situation involves filling in the cells in such a matrix with the frequencies of the relative events. For example, the relationship between pressing a light switch and a light coming on may be represented by a large number of pairings of the switch and the light (cell a) and a large number of occurrences of neither the switch nor the light (cell d). Unless the wiring is faulty, cells b and c would have entries of zero.

For binary variables, a variety of measures of contingency is possible (see Hays, 1963), but statisticians typically define covariation in terms of χ^2, which is defined as:

$$\chi^2 = N(ad-bc)^2/[(a+b)(c+d)(a+c)(b+d)],$$

where a, b, c, and d are the cell frequencies and N ($= a+b+c+d$) is the total number of events. In the last chapter I drew a distinction between structural and causal prediction, with the structural type involving predicting one attribute of a stimulus on the basis of others, and the causal type involving

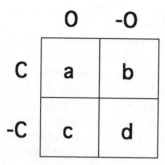

Figure 2.1. A contingency table representing the four possible combinations of events consisting of the presence and absence of a target cue C and an outcome O. Cell a is the frequency of conjunctions of the cue and outcome, cell b the frequency of occurrences of the cue without the outcome, cell c the frequency of occurrences of the outcome without the cue, and cell d the frequency of occasions on which neither event is present. –C and –O refer to the absence of the cue and outcome, respectively.

predicting an effect on the basis of a temporally prior cause. The χ^2 statistic is probably appropriate for structural prediction, but is less so from the point of view of causal prediction – which has been much more extensively studied – since it is a measure of the two-way dependency between a pair of events. Applied to a causal cue-outcome situation, it will yield large positive values if the outcome is dependent on the cue or if the cue is dependent (in the statistical sense) on the outcome. Since in causal prediction we are interested only in the question of whether the outcome is dependent on the cue, a slightly different measure is needed. Specifically, Allan (1980) has suggested that in this case the appropriate measure for a one-way dependency is the statistic ΔP:

$$\Delta P = P(O/C)-P(O/-C)$$
$$= a/(a+b)-c/(c+d)$$
$$= (ad-bc)/[(a+b)(c+d)],$$

where $P(O/C)$ is the probability of the outcome given the cue and $P(O/-C)$ is the probability of the outcome in the absence of the cue. Intuitively, there is no significant covariation between two events if the outcome is just as likely when the cue is present as when it is absent. Accordingly, ΔP is zero whenever the two conditional probabilities are zero. It approaches 1.0 as the presence of the cue tends to increase the likelihood of the outcome, and approaches –1.0 as the cue tends to decrease the likelihood of the outcome. In the latter case, we would still say that there is a degree of contingency or covariation between the events, but that the two events are negatively associated. In situations where $P(O/C)$ is greater than $P(O/-C)$, the cue to some degree causes the outcome, whereas in situations where $P(O/C)$ is less than $P(O/-C)$, the cue is a preventive cause which makes the outcome *less* likely to occur.

In the last three decades there has been a wealth of evidence concerning the effect on learning of varying the degree of contingency between the cue or action and outcome. To the extent that associative learning is appropriately influenced by degree of contingency, we can conclude that it is normative. Originally, results were decidedly mixed. In the 1960s, two influential studies suggested that people are at worst insensitive to contingency or at best only marginally influenced by it. Smedslund (1963), for example, presented student nurses with a sequence of 100 cards each of which reported the presence of a set of symptoms and diagnoses for a certain patient. The task was to estimate the extent to which symptom A was associated with diagnosis F, and in different conditions this contingency varied from positive to negative. Smedslund reported extensive over- and under-estimation of contingency, although subjects did seem to be at least partly influenced by it.

A more pessimistic conclusion – that subjects are entirely insensitive to variations in contingency – was reported by Jenkins and Ward (1965). They gave subjects 60 trials on which the action of pressing one of two buttons (A1 or A2) was followed by one of two outcomes (O_1 or O_2). The action–outcome contingency was the major independent variable, but the overall frequency of the outcome was also manipulated independently of contingency. At the end of each set of trials subjects rated their degree of control over the outcomes. Jenkins and Ward defined the actual degree of contingency as the difference between the conditional probability of the outcome given one response and the conditional probability given the other response. The results of the experiments showed that subjects' judgements deviated highly from the programmed contingencies, but were related to the overall frequency of the outcome. In fact judgements were unrelated to the actual contingency even when considerable efforts were made to remedy the insensitivity by giving subjects practice problems with feedback concerning the actual contingency.

In contrast to these early findings from humans, by the late 1960s sensitivity to contingency had been clearly established by Rescorla (1968) in conditioning experiments with laboratory animals, and soon the effect was established beyond question in humans. For example, Alloy and Abramson (1979) observed considerable sensitivity to event contingency in their subjects in a situation in which there was just one action and one outcome. In their studies, each set consisted of 40 trials on each of which the subject could choose whether or not to press a button, and a light either came on or did not. The subjects were required, at the end of the set, to rate the degree of contingency between responding and the onset of the light. The independent variable, actual contingency as measured by ΔP, varied from 0 to 0.75.

The crucial result of the experiments was that normal subjects accurately judged the degree of contingency in cases where ΔP was positive and in which the outcome, the onset of the green light, was affectively neutral. However, they overestimated the contingency in noncontingent ($\Delta P = 0.0$)

Table 2.1. *Design and results of the Shanks (1991a, Experiment 1) study*

Condition	Trial types	Test symptom	ΔP	Mean rating
Contingent	AB→O_1 B→no O	A	1.0	62.3
Noncontingent	CD→O_2 D→O_2	C	0.0	41.8

A–D are the cues (symptoms) and O_1, O_2 are the outcomes (diseases); no O indicates no outcome.

sets of trials when the outcome occurred frequently or was not neutral in the sense that it was related to monetary gain. They underestimated the contingency in contingent sets when the outcome was affectively negative, i.e. related to monetary loss. Depressed subjects, on the other hand, were highly accurate in their judgments. The conclusion reached by Alloy and Abramson was that people are indeed sensitive to contingency, but that certain situations lead to systematic biases. Finally, Alloy and Abramson were able to give a convincing explanation of why earlier studies had obtained little evidence of sensitivity to correlation. They argued that Jenkins and Ward's (1965) studies were difficult to interpret because subjects had been required to make one of two responses on each trial rather than to respond or not respond. Instead of contrasting one response with the other, if Jenkins and Ward's subjects were rating the degree of contingency between *either* response and the outcome, then insensitivity to the manipulated contingencies would have been observed.

Sensitivity to variations in contingency has also been shown to emerge in cue–outcome learning tasks. Table 2.1 shows the design and results of an experiment I conducted (Shanks, 1991a, Experiment 1) in which subjects saw hypothetical patients with certain symptoms (the cues). For each patient, the subject had to diagnose what illness (the outcome) they thought that patient had, and feedback was provided concerning the diagnosis. Some patients had symptoms A and B, and the correct disease was disease 1 (AB→O_1) and others had just symptom B and no disease (B→no O). Thus there is a positive degree of contingency between symptom A and disease 1, and this was reflected in subjects' judgments of the A→O_1 relationship (mean = 62.3 on a scale from 0 to 100). Some patients had symptoms C and D and disease 2 (CD→O_2) and others symptom D and disease 2 (D→O_2). The D→O_2 trials increase the probability of disease 2 in the absence of cue C, relative to the situation for cue A, and hence reduce the C→O_2 contingency, a fact that was reflected in the lower judgments for the C→O_2 rela-

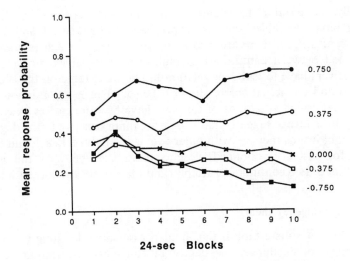

Figure 2.2. Mean response probability per sec across 10 blocks of 24 s under various action–outcome contingencies. The contingency values (ΔP) are shown to the right of each curve. Under positive contingencies response rate increased across trials while under negative contingencies it decreased. (After Chatlosh *et al.* 1985).

tionship than for the A→O$_1$ relationship (mean = 41.8). In the terminology of animal conditioning, cue D is said to have 'blocked' learning about the relationship between cue C and disease 2.

There is evidence that sensitivity to contingency is found when the dependent measure is response rate rather than a numerical judgment of causality. Figure 2.2 reproduces the data from a study by Chatlosh, Neunaber, and Wasserman (1985) in which subjects pressed a key in order to make a light flash, with each flash of the light earning the subject a point. The schedule was divided into 1-second (s) time intervals with P(O/A) and P(O/–A) determined for those intervals. The figure shows the mean probability of a response collapsed across 24-s blocks of trials for conditions in which ΔP was 0.750, 0.375, 0.0, –0.375, or –0.750. Chatlosh *et al.* found that response rates attained a higher level when there was a positive action–outcome contingency than when there was a zero contingency, which in turn yielded a higher rate than when there was a negative contingency. As can be seen from Figure 2.2, by the end of 4 min exposure to the schedules, there was a perfect ordering of the conditions with respect to the response probabilities they produced. Note also that the conditions were separated even on the first block, indicating that 24 s is sufficient to distinguish quite finely between different levels of contingency. Given this degree of sensitivity, it seems plausible that with sufficient motivation, subjects' response rates would have continued to climb under all positive contingencies to the point where a response was emitted on every trial, and would have declined to zero under all negative and zero contingencies.

Finally, in addition to situations in which relationships are learned between binary variables, such as the pressing of a light switch and the illumination of a light, there are also many situations in which relationships are learned between continuous or multi-valued variables. For instance, novice drivers have to learn the relationship between force on the accelerator pedal and the actual acceleration of the car. This is a continuous relationship. Although situations of this sort have been much less extensively studied, the evidence suggests that people are quite sensitive to the degree of covariation, just as they are with binary-valued variables. Beach and Scopp (1966), for instance, obtained close concordances between the judged covariation of two continuous variables and their statistical correlation, r.

Contingency theory

Plainly, these results establish that human associative learning is sensitive to, amongst other things, the degree of statistical covariation between events. Capitalising on this fact, there has been a long tradition – going back to the classic 1967 article by Peterson and Beach entitled 'Man as an intuitive statistician' – of arguing that human behaviour in learning tasks is not only correlated with normative measures such as r, χ^2, or ΔP, but is to all intents and purposes perfectly calibrated with such statistics. The outcome of comparisons between associative judgments and normative measures should tell us whether humans behave rationally or whether they are prone to systematic biases. More importantly, such comparisons allow us to determine what it is at the most general level that the learning system is doing. If it turns out that people's behaviour is closely matched to the contingency metric ΔP, then we can conclude that the system is in one way or another managing to compute the degree of contingency between events.

However, it is one thing to claim that judgments vary with variations in a certain statistic: it is quite another to suggest that performance is perfectly calibrated with respect to that statistic. Associative learning may in some circumstances correlate with ΔP, but do subjects behave *exactly* as prescribed by the ΔP formula? And if not, then is there an alternative non-normative statistic based on the cell entries of the contingency table which better describes performance? In this section, we will consider the evidence for and against the idea that associative learning conforms to the norm provided by contingency theory.

Although the ΔP rule is normative in the sense that it can be justified on statistical grounds, whereas other rules may not be justifiable, it is perfectly possible that subjects do fall back on simpler rules, and the fact that judgments tend to conform to ΔP does not exclude these alternative rules since the predictions of the various rules may be highly correlated. What is needed is some procedure to differentiate amongst them. As an illustration, consider an experiment by Allan and Jenkins (1983). They presented subjects with a cue–outcome learning task which took place on a computer. An image of a joystick on the left of the screen moved on some trials to indicate the cue, and the outcome consisted of a dot on the right-hand side of the screen moving downwards. Each of 50 trials was signalled by a band appearing on the screen. The band remained on for 1.3 s, during which the cue either occurred or did not, followed on some trials by the outcome. In order to make the predictions of the different rules as divergent as possible, the problems were run with different overall probabilities of the cue, P(C). This probability was either 0.5 or 0.7. Subjects were required to rate on a scale from 0 to 40 the influence of the cue on the outcome.

Allan and Jenkins programmed the 10 different problems shown in Table 2.2 in each of which the contingency between the cue and outcome, as determined by the values of P(O/C) and P(O/–C), was varied. For five of the problems, the objective covariation as measured by ΔP was zero, while in the remaining problems it varied up to 0.8. Each problem can be referred to by two numbers (e.g. 0.9/0.9) where the first refers to P(O/C) and the second to P(O/–C). To begin with, note that the five noncontingent problems did not yield equal ratings, as would be required if subjects were computing ΔP. Instead, ratings in these problems increased with the overall probability of the outcome (in Table 2.2 the ratings have been converted to a scale from 0 to 100). For the problems in which a covariation did exist, judgments tended to increase with increases in ΔP, but again were far from accurate. For instance, the two problems in which ΔP was 0.4 received very different ratings.

Of course, there are many potential rules that could be constructed using the cell entries of the contingency matrix, but there are six in particular that have attracted attention. Note that (apart from ΔP) these rules are all non-normative in the sense that they would not yield statistically-valid measures of relatedness. The 'cell a' rule assumes that judgments correspond to the number of pairings of the cue and outcome, in other words to the entry in cell a of the contingency matrix. Judgments describable by such a rule would be linearly related to the frequency of these trial types, and naturally, such a rule would in many cases yield highly inaccurate judgments. Nevertheless, it may still yield judgments that conform to Allan and Jenkins's data. The 'a–b' rule bases judgments on the frequency of the cue with (a) versus without (b) the outcome. The 'a+c' rule relates judgments to the total number of outcomes (cells a and c). In the 'ΔF' rule judgments are also dependent just on cells a and c, but are assumed to be related to the difference between the frequency of trials where the outcome occurs with the cue (a) and trials where the outcome occurs in the absence of the cue (c): that is, $\Delta F = a - c$.

The two most sophisticated rules are ΔD and ΔP, and these take all four cells of the contingency table into account. Clearly, cell a confirms the existence of a relationship, and equally, occurrences of the cue without the out-

Table 2.2. *Results of Allan and Jenkins' (1983) experiment*

P(O/C)	P(O/–C)	ΔP	P(C)=0.5			P(C)=0.7		
			ΔF	ΔD	J	ΔF	ΔD	J
0.1	0.1	0.0	0	0	**5**	2	–16	**13**
0.3	0.3	0.0	0	0	**14**	6	–8	**18**
0.5	0.5	0.0	0	0	**15**	10	0	**19**
0.7	0.7	0.0	0	0	**20**	14	8	**29**
0.9	0.9	0.0	0	0	**30**	18	16	**60**
0.3	0.1	0.2	5	10	**25**	9	–2	**13**
0.9	0.7	0.2	5	10	**28**	21	22	**43**
0.5	0.1	0.4	10	20	**33**	16	12	**38**
0.9	0.5	0.4	10	20	**48**	24	28	**69**
0.9	0.1	0.8	20	40	**70**	30	40	**75**

P(O/C) = probability of outcome given cue. P(O/–C) = probability of outcome in absence of cue. P(C) = probability of the cue. Judgments (J) have been converted to a scale from 0 to 100.

come (b) and the outcome without the cue (c) suggest that the cue is not positively related to the outcome. The ΔD rule proposes that judgments are a function of the difference between the evidence confirming a relationship and that disconfirming it:

$$\Delta D = (a+d)-(b+c).$$

For the purposes of this rule, cell d is regarded as confirmatory rather than disconfirmatory, although in reality trials on which both cue and outcome are absent are neutral with respect to the existence of a causal relationship.

At the general level, it does not seem as if any of the rules provides an especially good fit to the data. Subjects' judgments certainly do not correlate perfectly with ΔP, otherwise they would be equal in all of the noncontingent conditions, yet this is clearly not the case: judgments were very much higher in condition 0.9/0.9 than in condition 0.1/0.1, for instance. It is easy as well to dismiss the other rules. The 'cell a' rule predicts equal judgments in all of the conditions where P(O/C) is 0.9, since the number of cue-outcome pairings is equal in these conditions. Again, this is obviously not borne out by the data (and the 'a–b' rule fails for the same reason). The 'a+c' rule incorrectly predicts equal judgments in the 0.9/0.1 and 0.5/0.5 conditions where P(C) = 0.5, because in these conditions the total number of outcomes (a+c) is the same. Finally, as Table 2.2 illustrates, the ΔF and ΔD rules predict equal ratings in the noncontingent problems in the conditions where P(C) = 0.5 (and the same is true of the 'a–c' rule). However, assuming some appropriate monotonic function for mapping the obtained values of ΔF (or ΔD) onto judgments, these rules do fare better than the ΔP rule in that they can at least predict reasonably well the ratings observed when P(C) = 0.7. Across all problems, the median correlation in Allan and Jenkins's experiment between judgments and predictions was 0.73 for the ΔD rule, 0.65 for the ΔF rule, and only 0.50 for the ΔP rule.

These results are obviously discouraging for the view that associative learning is normative. However, it is possible that different subjects' judgments may conform to different rules and that by collapsing across subjects, regularities in the data are being obscured. For this reason, Allan and Jenkins correlated for each subject the judgments they gave under the different contingency conditions with the predictions from each of the rules. Unfortunately, the results confirmed the conclusions from the overall analysis: most subjects appeared if anything to be basing their judgments on the ΔD rule. In sum, judgments do not seem to conform to those predicted by the normative ΔP rule.

Rather contrasting results were obtained by Wasserman, Chatlosh and Neunaber (1983) in an action–outcome learning study. Subjects in this experiment pressed a telegraph key (the action, A) and judged the extent to which it caused a light to flash. The actual schedule consisted of dividing the 4 min duration of each condition into 1 s time intervals. If the subject responded during a given interval by pressing the telegraph key, then the white light flashed for 0.1 s at the end of that interval with a probability P(O/A), and if the subject did not respond during the 1 s interval, the light flashed with probability P(O/–A). At the end of each problem, subjects rated the action–outcome relation on a scale from –100 ('prevents the light from occurring') to +100 ('causes the light to occur'). Nine conditions were constructed by having P(O/A) and P(O/–A) take all combinations of the values 0.875, 0.500, or 0.125 per second.

Like Allan and Jenkins, Wasserman *et al.* found very little evidence to suggest that their subjects' judgments might conform to the 'cell a' or 'a+c' rules, which is perhaps not surprising since these rules correlate poorly with the objective contingency. For instance, we would expect that raising P(O/–A) will tend to reduce judgments, yet this has no effect on the number of action–outcome co-occurrences (cell a), and actually increases the total number of outcomes (cells a+c). However, for the ΔF, ΔD, and ΔP rules, substantial proportions of subjects' judgments were correlated with the predictions derived from each rule. Of course, a particular subject's judgments could be significantly correlated with more than one rule. But in general the ΔP rule came out best: in four subsets of subjects who differed in terms of their overall response rate, the mean percentage of subjects whose judgments were significantly correlated with the ΔP rule was 80%. The corresponding figure for the ΔD rule was 68%, and for the ΔF rule was 60%. Thus Wasserman *et al.*'s results are considerably more encouraging than were Allan and Jenkins's for the normative view.

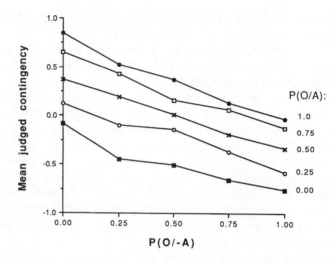

Figure 2.3. Mean judgments of contingency in 25 conditions formed by crossing different values of P(O/A) and P(O/–A). The abscissa shows P(O/–A), and points with the same value of P(O/A) are connected. Perfect conformity between judgments and ΔP would yield a set of parallel straight lines separated vertically by 0.25 units and with slopes of –0.5. (After Wasserman *et al.* 1993.)

In a further experiment using the same procedure, Wasserman *et al.* (1993) presented subjects with 25 different problems constructed by taking all possible pairings of P(O/A) and P(O/–A) with the values of 1.0, 0.75, 0.5, 0.25, and 0.0 per second. Again, the results indicated an impressive degree of sensitivity to the actual degree of contingency. Figure 2.3 shows the mean judgment given for each condition. Perfect concordance between judged and actual contingencies would result in a set of straight parallel lines with slopes of –0.5, with each pair of adjacent lines separated vertically by 0.25 points. Although not exactly conforming to this pattern, the correspondence is still excellent, so much so that 96.7% of the variance in the judgments is accounted for by the actual contingency ΔP. As P(O/A) was held constant, judgments decreased as P(O/–A) was raised from 0.0 to 1.0. Conversely, judgments increased (became less negative) when P(O/–A) was held constant as P(O/A) was raised from 0.0 to 1.0. Quite unlike the results obtained by Allan and Jenkins, judgments were close to zero when P(O/A) and P(O/–A) were equal. In sum, in Wasserman *et al.*'s procedure subjects were to a very high degree sensitive to the actual covariation between the action and outcome when judging their relatedness.

What are we to make of the different conclusions reached by Allan and Jenkins and by Wasserman and his colleagues? There are, of course, a number of differences between the tasks used. For example, Allan and Jenkins's subjects were making cue–outcome ratings while Wasserman *et al.*'s were making action–outcome ratings. However, the likeliest candidate is the rate of learning in each task. With a free-operant procedure such as that used by Wasserman and his colleagues, learning tends to be quite rapid, whereas in the discrete-trial procedure of Allan and Jenkins's study it is likely to be rather slower. This suggests that after further exposure, subjects in Allan and Jenkins's study might have been able to give ratings that accorded better with the ΔP rule. However, with the briefer amount of exposure that occurred in the actual experiment, judgments deviated substantially from the predictions of the rule. Clearer evidence comes from studies directly examining the evolution of associative ratings.

Learning functions

Several studies have looked at the way in which associative judgments progress as more experience of a contingency is acquired, and the results tend to support the above analysis. Typically, it has been observed that under a positive contingency judgments appear to increase as more experience of the contingency is provided. Judgments start close to zero and rise towards an asymptote at about the actual contingency. With a negative contingency judgments decrease from close to zero towards a negative asymptote. Figure 2.4 shows the learning curves obtained under a number of contingencies in an experiment by Francisco Lopez and myself. In that study, subjects saw tanks traverse a computer screen and either blow up or not. Subjects were told the tanks were passing through a minefield which contained colour-sensitive mines, and their task was to judge the relationship between the colour of a tank and whether it avoided being blown up. On each trial, a tank crossed the computer screen and was either coloured or not. Thus the cue (C) is the colour of the tank and the outcome (O) is avoiding being destroyed. The tanks were coloured on half the trials and not coloured on the remainder.

The experiment included four different cue–outcome contingencies, each lasting for 40 trials, with each subject seeing just one contingency. In one problem (0.75/0.25) the contingency was positive in that P(O/C) was greater than P(O/–C). In two problems (0.75/0.75 and 0.25/0.25) ΔP was zero but the overall probability of the outcome varied. Finally, in one problem (0.25/0.75) the cue and outcome were negatively related. Subjects made judgments every five trials on a rating scale from –100 to +100. Figure 2.4 shows these judgments, and as we would expect on the assumption that the difference between Allan and Jenkins and Wasserman *et al.*'s results was due to differences in exposure, judgments approached the actual values of ΔP as training proceeded. Judgments increased towards a value of 50 (= 0.5×100) in the 0.75/0.25 condition and decreased across trials towards –50 (= –0.5×100) in the 0.25/0.75 condition. In the noncontingent problems, judgments quite quickly converged towards the actual contingency of zero.

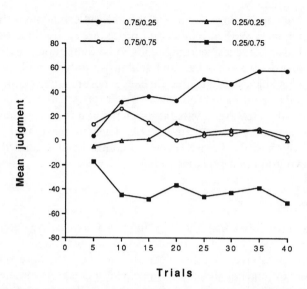

Figure 2.4. Mean judgments of contingency across 40 trials under four different cue–outcome contingencies. Judgments were made on a rating scale from –100 to +100. Each condition is designated by two numbers, the first being P(O/C) and the second P(O/–C). Judgments increase under the positive (0.75/0.25) contingency and decrease under the negative (0.25/0.75) contingency, in each case yielding terminal judgments close to the actual contingencies (×100). In the noncontingent conditions (0.75/0.75 and 0.25/0.25), judgments converge to zero, but when the probability of the outcome is high (in the 0.75/0.75 condition), early judgments are erroneously positive. (After Lopez & Shanks, unpublished data.)

Note, however, that after about 10 trials judgments in the noncontingent conditions were strongly biased, as in Allan and Jenkins's experiment, by the overall probability of the outcome, with judgments in the 0.75/0.75 condition being noticeably higher than those in the 0.25/0.25 condition. Nevertheless, the results point to the conclusion that associative judgments may approximate ΔP at asymptote, but deviate considerably prior to asymptote.

It is clear that although terminal judgments can approach the actual contingencies, the ΔP statistic does not accurately describe the shapes of these learning curves. As it stands, the ΔP rule in fact makes quite clear predictions about what the shapes of these acquisition curves should be: flat. This is because increasing the amount of exposure to P(O/C) and P(O/–C), the terms in the rule, does not change their mean values. Of course, more exposure will ensure that these probabilities are based on a larger sample of trials, and this will reduce the variance across trials in a group of subjects. But it will not affect the mean values of the probabilities. Plainly, this is in contrast with the subjects' judgments, which change dramatically across trials. In actual fact, for each block of eight trials in the experiment, the value of ΔP was exactly equal to its programmed value.

These results suggest two alternative conclusions concerning the normativity of associative learning. One is that judgments are normative at asymptote but biased prior to asymptote, and the other is that the ΔP metric is appropriate as an asymptotic but not as a pre-asymptotic norm. At present it is hard to know how to decide between these possibilities, but it is worth mentioning that alternative attempts to formulate normative theories, constructed for example around Bayes' theorem (see Fales and Wasserman, 1992), would predict learning functions that start at zero and slowly increase or decrease towards asymptote. Certainly, one could argue that it is rational for judgments to start close to zero: in the absence of any prior knowledge, the likelihood is that a pair of events that have never been seen before will be unrelated.

Note that the reason that Allan and Jenkins's data seemed to be in accordance with the ΔD rule is that the rule correctly predicts increasing and decreasing judgments under positive and negative contingencies, respectively. The terms in the ΔD rule $[\Delta D = (a+d)-(b+c)]$ are all zero at the outset. Under a positive contingency, the difference between $(a+d)$ and $(b+c)$ becomes steadily greater and hence ΔD increases as more trials are witnessed, while under a negative contingency ΔD steadily becomes more negative. Thus judgments may conform better with the predictions of this rule than with those of the ΔP rule simply because in failing to predict increasing judgments, the ΔP rule is doing particularly badly. The ΔD rule is unlikely to be a good model of the subjects' judgments, of course, because it fared poorly with Allan and Jenkins's noncontingent data and also because it predicts ever-increasing judgments under a positive contingency and ever-decreasing ones under a negative contingency, rather than judgments that level off.

The conclusion from this discussion is that, to a first approximation, associative judgments are unbiased at asymptote. When given sufficient exposure to a relationship, judgments match quite closely the contingency specified by the normative ΔP theory. However, substantial biases may occur prior to asymptote, with judgments commencing close to zero and only slowly regressing towards the predicted values. As we will see in Chapter 4, this regression process is a property of certain associationist learning mechanisms which can be said, in their asymptotic behaviour, to be computing ΔP.

Extinction and latent inhibition

In addition to the effect on learning of variations in contingency, there are some other simple and well-known effects that have a straightforward normative interpretation. Extinction is one such phenomenon. If a cue such as a tone is paired with an outcome like shock over several trials, and then is presented for several trials without the shock, then conditioned responding established in the first stage gradually fades until completely abolished, and judgments of the strength of the cue-outcome relationship also decline (e.g., Davey, 1992a). From a normative point of view, the unreinforced trials in the second stage serve to reduce the probability of the outcome given the cue, P(outcome/cue), and hence reduce ΔP.

There is one interesting exception to the general rule that repeated unreinforced presentation of a stimulus leads to a reduction in its associative strength. Studies of animal conditioning have established that repeated presentation of a cue that has been negatively associated with an outcome (a so-called 'conditioned inhibitor') does not lead to any change in its learned properties. Thus Zimmer-Hart and Rescorla (1974) presented animals with trials on which a light predicted shock (light→shock) and other trials on which a compound consisting of the light and a tone predicted no shock (tone+light→no shock). These trial types established a negative contingency between the tone and shock. At the end of the learning phase, the animals suppressed their ongoing activity in the presence of the light but not in the presence of the tone–light compound, indicating that they regarded the tone as being negatively related to shock.

The interesting result was that repeated presentation of the tone after the initial learning trials did not affect its ability to suppress responding to the light. Although this experiment has not been conducted with humans, it has been replicated so many times with animals that we can be fairly confident of its generality. Interestingly, a moment's reflection shows that this result is entirely consistent with contingency theory. During the training trials, the animals were able to learn that P(shock/tone) was zero whereas P(shock/no tone) was considerably greater than zero. Subsequent trials on which the tone was presented on its own merely confirm the fact that P(shock/tone) is zero and do not alter the computed value of ΔP. Thus it is quite amenable to contingency theory that repeated presentation of a stimulus that has a negative contingency for an outcome has no detectable consequences.

The second learning effect we shall consider is a phenomenon – again well-known in animal learning but only recently studied in humans – which from an informational point of view is identical to extinction except for the order in which the trial types are presented. If a cue is presented on its own without any consequences, then later pairings of the cue and an outcome will lead to a retardation in the rate of learning compared to a situation in which the cue has not been pre-exposed. This effect, called latent inhibition, has been demonstrated numerous times in conditioning studies and has had a considerable impact on the development of theories of conditioning (see Hall, 1991; Lubow, 1989). As an example, Lipp, Siddle and Vaitl (1991) conducted a human Pavlovian conditioning experiment in which the cue is the CS and the outcome is the US. Subjects in the experimental groups received paired presentations of a light CS and a tone US, and had to respond to the US by pressing a button as fast as possible. Learning was indexed by skin conductance during the cue. In one of these groups, there were no CS preexposures and substantial conditioning occurred to the light. In two other groups, the light was presented on its own for 10 or 20 trials, respectively, prior to the conditioning phase. In these groups, no conditioning was obtained. Thus preexposure to a cue in the absence of the outcome impairs later cue–outcome learning.

Like extinction, latent inhibition has a very straightforward normative explanation. Suppose that associative learning is related to the degree of contingency between the cue and outcome. In the control condition, $\Delta P = 1.0$ since the outcome occurs every time the cue occurs, but never in its absence. In the pre-exposure conditions, ΔP is less than 1.0 because it is not the case that the outcome occurs every time the cue is presented; the pre-exposure trials, just like a set of extinction trials, serve to reduce P(outcome/cue). Although the basic latent inhibition result can therefore be explained, I should mention that a potential problem for this account arises with respect to learning negative contingencies. Because the pre-exposure trials tend to reduce the estimated value of P(outcome/cue) across the whole experiment, we would have to predict that stimulus pre-exposure should increase rather than decrease the speed of learning a negative relationship. Although this has not been studied in humans, evidence from animals suggests that learning *both* positive and negative contingencies is retarded by pre-exposure (Reiss and Wagner, 1972).

Preparedness and phobias

One of the striking recent successes of associative learning theory has been the fact that therapies based on conditioning models of human phobias have proven highly successful. The conditioning perspective on phobias views them as fear responses conditioned to stimuli such as open spaces, dentists' waiting rooms, or spiders. If a particularly aversive event happens to a person in such a context, then any cues present at that time may come to be associated with the event and may acquire the ability to evoke anxiety. For instance, a particularly traumatic experience with a dentist may lead to fear being conditioned to the stimuli of the dentist's room. In order to avoid such anxiety, the person goes to considerable lengths to avoid the eliciting events: he or she avoids dentists, for instance. From the traditional associative learning perspective, simple techniques such as extinction can be used to reduce or eliminate the connection between the phobic stimulus and anxiety, and the evidence suggests a considerable degree of success in therapeutic techniques based on learning theory (Davey, 1992b).

My purpose in the present section is not to evaluate the conditioning approach to phobias, but rather to focus on one particular phenomenon concerning phobias which appears at first sight to represent a problem for the normative analysis of learning, and to try to see whether the normative analysis can be rescued. The phenomenon is so simple as to seem hardly worth mentioning: it is the fact that phobias tend to develop to some stimuli or objects and not to others. For instance, although spider phobias are very common, it would be rare indeed to find a person with a phobia for flowers. Now although visits to the dentist may objectively be more strongly associated with aversive outcomes (e.g. pain) than certain other activities, it seems very unlikely that the distribution of phobias is in general related to objective threat. For instance, aversive events probably do not occur in the presence of fear-relevant stimuli such as spiders any more often than in the presence of fear-irrelevant stimuli such as flowers: in fact, spiders are rarely signals of objective threat at all. Thus the unequal distribution of phobias is a problem for a normative theory.

One explanation is to say that certain events are predisposed to be easy to associate with aversive events and some are not. The standard form of this classic 'preparedness' theory, originally proposed by Seligman (1971), assumes that during our evolutionary history, some things such as spiders and snakes have been particularly dangerous. As a result our species has developed a rapid tendency to attach fear reactions to these things. Although nowadays the sorts of household spiders to which phobias often develop are perfectly innocuous, the rapid acquisition of fear still occurs and this in turn leads to the common development of phobic reactions.

What is the evidence for a differential ability of cues to enter into associations with aversive outcomes? Although the evidence is controversial (see McNally, 1987), laboratory studies have found that fear-relevant and fear-irrelevant stimuli do indeed differ in terms of the ease with which they are associated with outcomes. Consider the results of a study by Tomarken, Mineka and Cook (1989). In these experiments, Tomarken *et al.* exposed subjects to a sequence of slides of fear-relevant objects (snakes or spiders) and fear-irrelevant stimuli (mushrooms or flowers). Each slide could be followed either by a shock, by a tone, or by nothing. Then subjects were asked to estimate the relationship between each type of stimulus and each type of outcome. In actual fact, the probability of each outcome given each type of slide was 0.33, so subjects should have given equal ratings for all relationships. But what they actually did was to dramatically overestimate the probability in one of the conditions, namely the one in which fear-relevant stimuli were associated with the aversive outcome, shock.

Tomarken *et al.*'s result is consistent with the idea that some stimuli are 'prepared' in the sense that they are particularly easy to associate with aversive consequences. Of course, if this theory is correct, then it would be inconsistent with the view that learning is simply governed by the objective degree of contingency between the cue and the outcome, because a given individual with a spider phobia is likely to have been exposed to no greater a contingency between spiders and aversive outcomes than between flowers and aversive outcomes (as indeed the Tomarken *et al.* experiment shows).

But it turns out that differences between fear-relevant and -irrelevant stimuli are not necessarily inconsistent with the normative model.

To see how this can be the case, consider the results of an experiment by Davey (1992a). Davey informed subjects that shocks might occur following some stimuli but not others, and then simply asked the subjects to rate the likelihood of an imminent shock during each slide presentation. Subjects gave consistently higher ratings when they were presented with fear-relevant rather than fear-irrelevant stimuli, and this bias occurred even on the very first trial. Such a result suggests that the difference between fear-relevant and fear-irrelevant stimuli comes about not because of biological preparedness for association with an aversive event, but rather because subjects come to the laboratory with an elevated expectation of an aversive event in the presence of a fear-relevant stimulus. That is to say, subjects commence the task with an inflated estimation of P(shock/spider) compared to P(shock/flower). That being the case, their subsequent behaviour – associating the fear-relevant but not the fear-irrelevant stimulus with the aversive outcome – is entirely consistent with the normative account.

Of course, it remains an open question as to where these inflated estimates come from. It is certainly possible that they have a biological origin, but what is much more likely, as Davey (1995) points out, is that they derive from cultural and social transmission. From childhood upwards, we are exposed to images and information priming us to treat spiders and snakes as potentially threatening objects. It would be little wonder if this information came to be represented in the mind of the average person as an elevated estimation of the probability of an aversive event in the presence of a spider. But whatever their origin, the differential effects seen with fear-relevant and fear-irrelevant stimuli do not seem especially problematic for a normative account of associative learning. We will return later to briefly consider some other examples of selective associations that may not be so readily explicable in these terms.

Judgmental accuracy

Before continuing the discussion of the ΔP theory, it is worthwhile briefly considering the general level of accuracy that subjects achieve when asked to judge associative relationships, and to ask exactly what the theoretical significance is of the level of judgmental inaccuracy, where by inaccuracy I mean deviation from ΔP. As we have seen, there are circumstances in which subjects can achieve impressive levels of accuracy as asymptote is approached. Wasserman *et al.*, for instance, found that instrumental ratings under a variety of action–outcome contingencies were extremely close to the actual programmed contingencies. However, there are also other situations in which the concordance between programmed contingencies and judgments is much lower. Does the *relative* inaccuracy of these judgments have any theoretical significance? Certainly, if another judgmental rule such as ΔD yields a closer correspondence with the observed judgments, then the inaccuracies should be regarded as negative evidence from the point of view of the normative theory. I shall argue, though, that unless this is the case, discrepancies should be treated with a good deal of caution, and strong theoretical inferences should not be drawn from the overall level of accuracy or inaccuracy.

Much of the data relevant to the question of the normativity of associative learning come from experiments in which the dependent variable is a judgment on a rating scale. Since these scales usually go from −100 to +100, and since the statistic ΔP goes from −1 to +1, it is often assumed that judgments are normative whenever they correspond closely to ΔP (×100) and are non-normative otherwise. However, this may be an unfair constraint to impose on the normative theory. Suppose that associative knowledge is represented by some internal state with a parameter representing the subject's belief about the strength of the relationship, and let us take the normative view that at asymptote the internal parameter will be very close to ΔP. The exact nature of the psychological function mapping this internal parameter into an actual judgment or into behaviour in a given experimental context is unknown, but the nature of this function is critical. Thus far we have implicitly assumed that the function simply multiplies the internal parameter by 100 to translate it onto the rating scale. However, the function may be more complex than this.

For instance, there is evidence that with some rating scales, a cue which apparently has an internal strength of zero can elicit substantially positive judgments, which implies that the mapping function is not a straightforward one. Some relevant evidence comes from another study by Wasserman in which subjects were asked to make contingency ratings for a number of different action–outcome contingencies (Neunaber and Wasserman, 1986). In five problems, the contingency was zero, with $P(O/A)$ and $P(O/-A)$ being equal and having the values 0.1, 0.3, 0.5, 0.7, and 0.9 per second in the different problems. For subjects who were allowed to use a bidirectional scale going from −100 to +100, with negative ratings corresponding to negative contingencies, ratings for the noncontingent problems were all very close to zero. On that basis, then, we would have to conclude that the internal parameter reflecting the action–outcome relationship had a value close to zero. In contrast, subjects who had to use a unidirectional scale from 0 to 100 gave mean ratings for these noncontingent problems that were markedly greater than zero. In fact, across all these conditions the mean was approximately 20.0. The clear conclusion is that the discrepancies between judged and actual contingencies that emerge when unidirectional scales are used are of little theoretical significance, since judgments do not necessarily correspond to the internal variable that represents the subject's knowledge. As Neunaber and Wasserman (1986, pp. 177–8) say,

> . . . when only the magnitude of the response–outcome relationship is assessed, the positive mean ratings of the noncontingent problems appear to reflect a measurement artifact, rather than true perceptions of a positive response–outcome relationship. These findings suggest considerable caution in interpreting the results of studies that use a unidirectional rating scale and report positive mean ratings of noncontingent problems . . .

Of course, use of a unidirectional rating scale is only one of many potential factors that may introduce distortions into the mapping from internal knowledge onto observed behaviour. Another possible factor is the payoff schedule in operation: a subject may not believe there is an action–outcome relationship, but the reward for obtaining the outcome may be so great and the cost of performing the action so small that response rate attains a high level. For instance, in an action–outcome learning experiment that Anthony Dickinson and I conducted (Shanks and Dickinson, 1991) and which was briefly discussed in the last chapter, two groups of subjects were exposed to the same noncontingent schedule but under different task demands. One group of subjects, instructed they could earn points each time the outcome occurred, pressed a key about 20 times per minute when the optimal thing to do was not respond at all. The other group, asked to judge the action–outcome contingency, gave estimates very close to zero. Thus the same action–outcome schedule can elicit inconsistent behaviours depending on the demands of the task. We need to bear in mind that discrepancies between the predictions of a given theory and observed behaviour may be attributable to a complex mapping from internal state to observed behaviour.

Moreover, what are we to do when the dependent variable is something other than a judgment? Suppose we are studying response rate, for instance, as our dependent variable, and we obtain response rates under a variety of contingencies just as Chatlosh *et al.* (1985) did in the experiment described earlier (Figure 2.2). Are these response rates 'normative'? It is not clear how one should answer that question, because all our normative theory states is the value of ΔP. If ΔP in a certain problem is, say, 0.5, then what response rate should we expect to see if subjects are behaving normatively? There is no obvious answer. Instead, all we can rely on is ordinal data: assuming that the payoff structure provides more reward for more occurrences of the outcome, an increase in contingency should lead to an increase in response rate. Thus when the dependent measure is something other than a numerical judgment, all that the normative theory says is that our dependent measure (e.g., response rate) should correlate perfectly across a series of conditions with ΔP. Such a correlation is evident, of course, in Chatlosh *et al.*'s data.

Selectional effects

Up to this point I have concentrated on situations in which subjects are required to judge the relationship between an action or cue and an out-

Table 2.3. *A hypothetical contingency table*

	outcome	no outcome
action	50	50
no action	50	50

For this table, $P(O/A) = P(O/-A) = P(A) = 0.5$ and $\Delta P = 0.0$. Cell entries are event frequencies.

come, and where the action or cue is the only obvious causal event present. For such situations, contingency is very easy to specify. However, it is rarely (if ever) the case that potentially predictive cues appear in isolation; instead, we are often confronted with a set of potential cues that co-occur with one another, and we have to pick out the one (or ones) that is (are) truly predictive. Accordingly, we need a way of applying the ΔP statistic in these more complex cases. In the present section we will consider a simple but powerful solution recently developed by Cheng and Holyoak (1995) called the 'probabilistic contrast model'. As before, we will ask whether judgments conform to the prescriptions of this normative theory when multiple causal cues are present.

Table 2.3 shows a hypothetical set of frequencies of each of the four combinations of an action and an outcome that might be observed if $P(O/A)$ and $P(O/-A)$ were both 0.5, and if $P(A)$ were also 0.5. According to our statistical measure of association, there is no relation between these events, since $\Delta P = 0.0$. In an experiment I conducted under these conditions (Shanks, 1989), subjects judged (on a scale from 0 to 100) the degree of contingency between pressing a key on a computer keyboard and a light flashing on the screen, with the conditional probabilities being defined as before over 1-s time intervals. The use of a unidirectional rating scale probably accounts for the fact that subjects overestimated the contingency when given 3 min to witness the relationship: as Figure 2.5 shows, the mean judgment was 20.9 in this 0.5/0.5 condition. In a contingent (0.5/0.0) condition with $P(O/A) = 0.5$ and $P(O/-A) = 0.0$, judgments were of course substantially greater.

The interesting result of the experiment comes from another condition that was identical to the noncontingent condition of Table 2.3 except in one respect. In this condition, all occurrences of the outcome in the absence of the action (cell c) were accompanied by a tone stimulus. Here, if the subject had not responded by the end of a given 1-s interval, then if the outcome was programmed to occur, it was delayed momentarily and immediately preceded by the tone which lasted for about half a second. Although

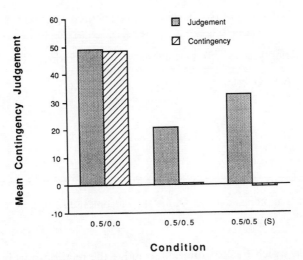

Figure 2.5. Effects of a cue signalling noncontingent outcomes. Judgments were made on a rating scale from 0 to 100, and each condition is designated by two numbers, the first being P(O/A) and the second P(O/–A). Mean judgments of contingency and mean actual contingency is shown for each of three conditions. In one condition (0.5/0.0), the outcome was contingent on the action, while in the second condition (0.5/0.5) the cell frequencies (shown in Table 2.3) produced a contingency of zero. In the third condition [0.5/0.5 (S)], every occurrence of the outcome in the absence of the action was signalled by a tone. The signal partially counteracted the effect of introducing noncontingent outcomes. (After Shanks, 1989.)

P(O/A) and P(O/–A) were again both 0.50, and therefore ΔP was again zero, Figure 2.5 shows that in this 'signal' condition, denoted 0.5/0.5(S), subjects gave a mean judgment that was significantly greater than that in the original 0.5/0.5 condition. According to the standard theory, although an additional predictive event has been introduced for all of the cell c events, ΔP between the action and outcome is identical in the two conditions. So why does the difference in judgments occur?

This result, which had earlier been demonstrated in animal conditioning experiments by Durlach (1983), is an apparent puzzle for a normative analysis such as the ΔP rule, because of course the introduction of additional events does not alter the contingency between the action and outcome. Note that at first glance it seems as if subjects are behaving quite irrationally in the signal condition: although their overestimate of the contingency may be partly interpreted as a result of the unidirectional rating scale, they believe more strongly that the action and the outcome are related than they do in the condition without the signal, which from an objective viewpoint is not the case.

Before considering how the normative theory might be revised to account for this result, let us first consider another example of the same problem. This comes from a cue–outcome learning experiment (Shanks, 1991a,

Table 2.4. *Design and results of the study by Shanks (1991a, Experiment 2)*

Condition	Trial types	Test symptom	Mean rating
Contingent	$AB \rightarrow O_1$ $B \rightarrow$ no O $C \rightarrow O_1$	A	58.6
Noncontingent	$DE \rightarrow O_2$ $E \rightarrow O_2$ $F \rightarrow$ no O	D	33.8

A–F are the cues (symptoms) and O_1, O_2 are the outcomes (diseases); no O indicates no outcome.

Experiment 2) in which I asked subjects to judge the relationships between symptoms and fictitious diseases. Each subject was presented with the six trials types shown in Table 2.4. At first glance, the design seems hardly different from that adopted in the earlier contingency experiment shown in Table 2.1: subjects received $AB \rightarrow O_1$ and $B \rightarrow$ no O trials together with $DE \rightarrow O_2$ and $E \rightarrow O_2$ trials, and these trial types establish a more positive contingency between A and O_1 than between D and O_2. The only difference is the addition of $C \rightarrow O_1$ and $F \rightarrow$ no O trials. Unsurprisingly, the result was that ratings of cues A and D differed, with the mean rating for A as before being greater than that for D. Yet, it turns out that in this experiment ΔP is equal for cues A and D – and hence the difference in judgments is unexplained – if contingency is computed across all of the trials. The addition of the $C \rightarrow O_1$ and $F \rightarrow$ no O trials radically changes the values of ΔP for each cue, since in this design, $P(O_1/A) = P(O_2/D)$, while $P(O_1/-A) = P(O_2/-D)$. Hence ΔP must be equal for cues A and D.

Results such as these have been interpreted as illustrations of *cue selection* in learning tasks, meaning that subjects apply some principles of selection to choose between potential causal factors. Without changing the objective contingency between a cue and an outcome, in some circumstances the cue may be selected for association with the outcome and in other circumstances not, depending on the presence and status of other cues. Such selection effects have been regarded in a number of different ways: at one extreme has been the view that they are signs of the fundamental irrationality of human learning, while at the other has been the view that they represent the near-perfect adaptiveness of the human learning mechanism to the environment. Our present purpose is to see whether these sorts of results are genuinely damaging for normative views. I shall refer to these effects as 'selectional', taking that term to be descriptive rather than theoretical.

In all of the simple cue– or action–outcome learning experiments considered previously, in which for instance subjects press a key and observe a light flashing, I have simply assumed that the terms that appear in the ΔP rule, P(outcome/cue present) and P(outcome/cue absent), are straightforward to determine so long as we have a specified time window or trial structure. With respect to the 2×2 contingency table, we assume that each occurrence of the outcome can be readily categorised as one occurring with or without the target cue or action. Each experimental trial either contains the cue (or action) or does not, and either ends with or without an occurrence of the outcome. However, from a causal point of view this scenario is somewhat odd, because an outcome occurring in the absence of the target cue must have been caused by something else: there must be some third event, ignored in the above analysis, that caused the outcome. Thus the very notion of varying the degree of contingency between a cue and outcome presupposes that outcomes may occur in the absence of the cue, which in turn requires that some third event must be present to have actually caused the outcome. In a typical experiment, such as those conducted by Wasserman and his colleagues, we must suppose that the experimental context acts as a background cue to which the subject may attribute outcomes. While this background is constant in simple experiments, in the signal condition of the Shanks (1989) experiment the background (which includes the signal) is not constant. The same is true of the experiment shown in Table 2.4.

How can these considerations be accommodated within a normative theory? Statisticians do not have any agreed procedure for specifying the contingency between events A and B when the background varies, but psychologists and philosophers have offered one fairly straightforward solution, developed in detail by Cheng and Novick (1990, 1992) and Salmon (1984), and applied directly to associative learning tasks by Cheng and Holyoak (1995). Cheng and Holyoak call it the 'probabilistic contrast model'. The procedure is to calculate P(O/C) for a target cue C across trials on which C occurs, and to calculate P(O/–C) across trials that are identical to the C trials with the exception that C is absent. That is to say, ΔP should be calculated not across all trials but rather across a subset (the 'focal set') of trials in which background events are kept constant. If we designate the background as B, then the calculation is

$$\Delta P = P(O/C.B) - P(O/-C.B), \tag{2.1}$$

where P(O/C.B) is the probability of the outcome given the cue and background, and P(O/–C.B) is the probability of the outcome in the absence of the target cue and presence of the background. Actually, a positive value of ΔP from this equation does not guarantee that the cue is the sole cause of the outcome, because an alternative interpretation is that the *conjunction* of cue and background is necessary. To rule this out, Cheng and Holyoak point out that it is necessary to calculate yet another form of contingency, namely:

$$\Delta P = P(O/C.-B)-P(O/-C.-B), \tag{2.2}$$

which is a measure of the cue–outcome contingency when B is absent (in other words, in another context). If both of these versions of ΔP yield positive values, then a reliable relationship exists. On the other hand, if Equation 2.1 yields a positive value of ΔP, but Equation 2.2 yields a value of zero, then the cue is only causal when it interacts with the background B. For instance, pressing a key on a computer keyboard only makes a light flash on the screen in the context of a specific experimental procedure.

Some concrete examples may be useful. Suppose we are trying to ascertain whether vitamin supplements are causally related to school achievement, and suppose that vitamin intake is correlated with a well-established cause of high achievement such as good overall nutrition. There are four possible scenarios. First, there may be a positive contingency between vitamins and achievement in well-nourished children, and a positive contingency in poorly-nourished children. In this case, vitamins will be interpreted as a genuine cause (unless taking vitamins is correlated with some other cause of achievement). Secondly, there may be a zero contingency between vitamins and achievement both for well- and for poorly-nourished children, in which case there is no predictive relationship between vitamins and achievement. Thirdly, if vitamins are contingently related to achievement in well-nourished but not poorly-nourished children, then vitamins will not be interpreted as being directly related to achievement: instead they will be viewed as interacting with overall nutrition to increase school achievement. Finally, if vitamins are unrelated to achievement in well-nourished children but are related to it in poorly-nourished children, then they will again be interpreted as interacting with overall nutrition level to affect school achievement. But in this case, the lack of contingency in well-nourished children is likely to be interpreted as a ceiling effect: these children are already doing as well as possible, and extra vitamins cannot improve their performance further.

As another example, consider the case of the relationship between the use of sun-tan lotion and getting sunburn, which intuition suggests should be a good example of a negative contingency. Suppose that like thousands of other British people, I forget to use sun-tan lotion and get sunburned, with distressing predictability, on the first day each year that the sun comes out. The next time the sun comes out, I remember to use sun-tan lotion, but even so, there is still a 50% chance that I get sunburned: perhaps I stay out so long that the sun-tan lotion fails to protect me. On the face of it, sun-tan lotion is negatively correlated with getting sunburn, but such a conclusion only follows if contingency is computed conditionally. If we compute unconditional probabilities, we find that the probability of getting sunburn in the absence of using sun-tan lotion is about 1/364: there is one day a year on which I use sun-tan lotion, and 364 on which I do not, and on one of these I get sunburned. The probability of getting sunburned on the one day that I use sun-tan lotion is 0.5. Hence the relationship between using sun-tan lotion and getting sunburned is positive:

$$\Delta P = P(\text{sunburn/sun-tan lotion})-P(\text{sunburn/no sun-tan lotion})$$
$$= 0.5-1/364$$
$$= 0.497.$$

Thus contrary to intuition, sun-tan lotion seems to predict the occurrence, and not the avoidance of sun-tan. But of course this conclusion comes about because we have computed contingency unconditionally: in a sense, using sun-tan lotion *does* predict sunburn, since I only use sun-tan when I am out in the sun. But it is not sun-tan lotion that is the explanatory factor, rather it is something correlated with it, namely the presence of the sun. Instead, what we need to do is compute contingency conditionally, which in this case means that we need to look at the probability of sunburn given that I use sun-tan lotion (which implies I am out in the sun) and the probability of sunburn given that I do not use sun-tan lotion, but where I am again out in the sun. With these probabilities, we will of course obtain a negative contingency. Thus our intuition that there is an inhibitory relationship between the use of sun-tan lotion and sunburn is correct, so long as we compute contingency conditionally.

It is important to note that Cheng and Holyoak's extension of contingency theory only changes our predictions when the background is variable. For all of the experiments considered in previous sections, the predictions are unaffected since the background is constant. The only addition is the perfectly sensible and valid point that since subjects in these earlier experiments could only have applied Equation 2.1 and not Equation 2.2 (because they were just exposed to one context), they are only licensed to conclude that the action or cue is an interactive cause when combined with the background. The key idea, of course, is that the evaluation of a cue must be based on a contrast between what happens when it is present versus what happens when it is absent, all else being held constant, and this idea should require little justification in the context of a scientific methodology which emphasises the use of controlled experiments that adopt exactly this procedure. In scientific experiments, the researcher sets up an experimental and a control condition in which everything is held constant other than the presence versus absence of the critical factor.

Cheng and Holyoak (1995) have argued not only that the probabilistic contrast model is the appropriate normative theory for causal or associative relationships, but also that human behaviour is closely matched to it. With respect to the signal effect (Figure 2.5), the theory says that ΔP is different in the 0.5/0.5 and 0.5/0.5(S) conditions. For the straightforward noncontingent condition, the theory's computation of ΔP is the same as before: the probability of the outcome given the action, P(O/A), and the probability of the outcome in the absence of the action, P(O/–A), are both 0.50. However, in the signal condition P(O/–A) must now be determined across trials that are identical to the action trials, except that the action is absent. The action trials are, naturally, action plus context trials, so P(O/–A) is calculated across context-only trials. But, the outcome never occurs on context-only trials, hence P(O/–A) is 0.0, and ΔP is 0.50. Another way of seeing this is to look again at Table 2.3. P(O/A) is calculated from the top two cells, cells a and b:

$$P(O/A) = a/(a+b) = 50/(50+50) = 0.50.$$

P(O/–A) is normally calculated from the bottom two cells, c and d. However, in the 0.5/0.5(S) case the events in cell c are removed because on these trials the signal is present, which means they are excluded from the focal set, and c is therefore zero. Thus:

$$P(O/–A) = 0/(0+50) = 0.0,$$

which means that ΔP is 0.50. The theory therefore produces a difference in the required direction between the value of ΔP in the noncontingent and signal conditions.

With regard to the experiment (Shanks, 1991a, Experiment 2) shown in Table 2.4, the procedure is to calculate $P(O_1/A)$ across all trials, which in this case means the AB trials, yielding a conditional probability of 1.0. For $P(O_1/–A)$, we focus only on trials that are identical to the trials that contribute to the computation of $P(O_1/A)$, except that A is absent; in this case, that means we consider only the B trials, from which we obtain $P(O_1/–A) = 0.0$ and hence ΔP = 1.0. A comparable computation for cue D yields a value of ΔP = 0.0, and hence the effect of contingency is accounted for. The idea is simply that the causal efficacy of a cue has to be determined by contrasting what happens when the cue is present versus what happens when it is absent, everything else being held constant. Note that we normally restrict the trials that contribute to the computation of ΔP anyway: we ignore trials occurring outside the experimental setting.

The probabilistic contrast model represents a major advance in our conception of normative theories of associative learning. Prior to its development, it had been widely assumed that learning was non-normative because of the sort of cue–selection result shown in Figure 2.5 and Table 2.4. Researchers assumed that P(O/C) and P(O/–C) had to be calculated unconditionally, that is to say across the entire set of experimental trials, and when computed in this way, judgments quite clearly do not correspond to ΔP. For instance, Chapman and Robbins (1990) concluded that data they obtained refuted contingency theory. They asked their subjects to make predictions about changes in a fictitious stock market, and the design is shown in Table 2.5. In the first part of the learning phase, whenever stock A rose in price, the market rose in value as well (the outcome, O). Whenever

Table 2.5. *Design and results of the experiment by Chapman and Robbins (1990)*

Stage 1	Stage 2	Test trials	Mean judgment
A→O	AC→O	C	31
B→no O	BD→O	D	77

A–D are the cues (stocks), O is the outcome (market increase), and no O indicates no change in the market.

stock B rose in price, the market failed to increase. Thus A was a good predictor of market increase while B was not. In the second part of the learning phase, on some trials stocks A and C increased together, and the market rose, and on other trials stocks B and D rose together and the market again rose. In the test phase subjects rated on a scale from –100 (perfect predictor of market not rising) to +100 (perfect predictor of market rising) how well each stock predicted a change in the market.

Note that stocks C and D are associated with a rise in the market on exactly the same number of occasions. Nevertheless, when asked to make predictive judgments for each stock, subjects gave a higher rating for stock D than for stock C. Thus here we have another example of cue selection, with stock A 'blocking' stock C. If we assume that ΔP is calculated unconditionally across *all* trial types, Chapman and Robbins's result cannot be explained because $P(O/C) = P(O/D)$ and $P(O/-C) = P(O/-D)$, meaning that ΔP is identical for D and C (it actually has the value 0.57 in this experiment).

However, when calculated across the appropriate focal sets of trials, rather than unconditionally, the picture is much more encouraging for contingency theory. If ΔP for cue C is calculated just across the AC and A trials, as the probabilistic contrast model suggests, and ΔP for cue D is calculated across the BD and B trials, we obtain $\Delta P = 0.0$ for cue C and $\Delta P = 1.0$ for cue D. The zero contingency for cue C comes about because the outcome has the same probability on AC and A trials:

$$\Delta P_C = P(O/C.A) - P(O/-C.A) = 1.0 - 1.0 = 0.0$$

where $P(O/C.A)$ is the probability of the outcome in the presence of both C and A and $P(O/-C.A)$ is the probability of the outcome in the presence of A and the absence of C. In contrast, the probability of the outcome differs on BD and B trials:

$$\Delta P_D = P(O/D.B) - P(O/-D.B) = 1.0 - 0.0 = 1.0.$$

Thus the theory predicts a difference in the right direction between judgments for C and D. Note that the observed difference is nothing like as extreme as expected, so this discrepancy would presumably have to be explained by saying, for instance, that the judgments have not reached asymptote. At any rate, it is clear that the selection effect Chapman and Robbins obtained can in principle be explained in terms of the computation of contingency.

A further demonstration of causal selection that can be explained in terms of Cheng and Holyoak's theory comes from an elegant study by Baker *et al.* (1993). They demonstrated that if an action, moderately well-correlated with the occurrence of an outcome, was accompanied by an alternative causal agent that was in fact a perfect predictor of the outcome, then judgments of the action–outcome relationship were substantially reduced. In their first experiment subjects were required to participate in a video game similar to that used in the experiment by Lopez and myself described earlier. The subjects were able to fire at tanks that passed across the video screen. By firing at the tanks, subjects could change their colour, which might help them avoid colour-sensitive mines in a minefield the tanks had to traverse. On some trials, then, the tanks were detected by the mines and were blown up. The subjects' task was to determine the extent to which the action of firing at the tanks caused the outcome of avoiding destruction. The basic design is shown in Table 2.6, which shows the possible outcomes on trials where the subject fired at the tank and on trials where they refrained from firing.

The contingency was such that the probability of the tank avoiding destruction given a hit by the subject was 0.75 while the probability given no hit was 0.25. Thus the contingency, ΔP, between firing and the tanks avoiding destruction was 0.50. In addition, though, there was another stimulus in the game (the signal), and this consisted of a plane which flew over the tanks on certain trials. Subjects were told that the plane was able to relay to the tanks information about the mines, and so make them less likely to be destroyed. In condition 0.5(0.0), the plane in fact had no effect: it was just as likely to appear on trials where the tank was not destroyed as on trials where it was, and so its contingency with tank destruction was zero (see second column of Table 2.6). Subjects' judgments in this condition were very close to 50 on a scale from 0 to 100, indicating that the plane probably had little interfering effect on judgments. However, in condition 0.5(1.0), the planes only appeared on trials where the tank avoided destruction, and hence their presence was perfectly correlated with the avoidance of destruction (right column of Table 2.6). The result was that action–outcome judgments in this condition were significantly, and very substantially, reduced even though the statistical relationship between firing at the tanks and them avoiding destruction was just the same as in condition 0.5(0.0). Hence this result, which Baker *et al.* obtained in a series of experiments with different contingencies and using different experimental procedures, stands as another clear illustration of interaction or selection between potential causal agents.

Table 2.6. *Design and results of Baker et al.'s (1993) experiment*

Trial types	Condition	
	0.5(0.0)	0.5(1.0)
A+S→O	7.5	15
A+S→no O	2.5	0
A+no S→O	7.5	0
A+no S→no O	2.5	5
no A+S→O	2.5	5
no A+S→no O	7.5	0
no A+no S→O	2.5	0
no A+no S→no O	7.5	15
P(O/A)	0.75	0.75
P(O/–A)	0.25	0.25
ΔP_A	0.50	0.50
P(O/S)	0.50	1.0
P(O/–S)	0.50	0.0
ΔP_S	0.0	1.0
Mean judgment	49	–6

The second column of the table shows the trial types (and frequencies) in condition 0.5(0.0) where the signal (S) was noncontingently related to the outcome (O). The third column shows the trial types in condition 0.5(1.0) where the signal was perfectly correlated with the outcome. A = action, no A = no action, no S = signal absent, no O = no outcome.

How does Cheng and Holyoak's theory explain this result? Equations 2.1 and 2.2 must be applied in order to yield a value of ΔP conditional on the presence of the signal and another value conditional on its absence. In condition 0.5(0.0), the action–outcome contingency when the signal is present, $\Delta P/S$, is:

$$\Delta P/S = P(O/A.S)-P(O/-A.S) = 0.75-0.25 = 0.50,$$

and the contingency when the signal is absent, $\Delta P/-S$, is the same:

$$\Delta P/-S = P(O/A.-S)-P(O/-A.-S) = 0.75-0.25 = 0.50.$$

As we saw earlier, whenever ΔP is positive both in the presence and in the absence of the conditionalising event (in this case, the signal), then a genuine relationship exists. Thus we would expect subjects to rate the action, as indeed they did, as positively related to the outcome.

Turning to the 0.5(1.0) condition, we find that in this case ΔP is zero whether or not the signal is present:

$$\Delta P/S = P(O/A.S) - P(O/-A.S) = 1.00 - 1.00 = 0.00,$$

when the signal is present and

$$\Delta P/-S = P(O/A.-S) - P(O/-A.-S) = 0.00 - 0.00 = 0.00$$

when it is absent. When both probabilities are zero, we can conclude unequivocally that no action–outcome relationship exists. As desired, the modified theory predicts a difference in judgments, with the 0.5(0.0) condition yielding a positive relationship and the 0.5(1.0) condition yielding a zero relationship, which is of course the observed result.

A summary of where we have got to may be in order. People appear to select amongst predictive cues: when the unconditional contingency between a target cue and an outcome is held constant, the extent to which the cue and outcome are associated depends on the status of other cues that are concurrently present. If these other cues are themselves highly predictive of the outcome, then the target cue will be to some extent ignored, while if the other cues are not especially informative, the target event will receive a normal association with the outcome. While such results are impossible to explain in terms of the computation of unconditional contingency – that is to say, in terms of the computation of ΔP across all trials in the experiment – they can be understood with reference to contingency calculated across subsets of the trials. Specifically, a focal set consists of all trials in which the target cue is present as well as those trials which are identical to the target-present trials except for the absence of the target. In this way, contingency is calculated by considering the difference in the probability of the outcome when everything is held constant except for the addition of the target cue.

Biases in learning

There is no doubt that the range of applicability of normative models based on contingency has been enormously extended by Cheng and Holyoak's analysis, and there is also little doubt that – within certain constraints – associative learning is very close to being normative. In fact, in the remainder of this book I shall follow Dickinson (1980), Anderson (1990) and many others in assuming that the associative learning mechanism has been shaped by evolution to detect statistical contingency. I shall also assume that, like visual illusions, deviations from the prescriptions of the normative theory are likely to be understood by examining the specific mechanism that underlies learning. One can either interpret illusions as evidence that perception is inherently biased, or as the result of a system that has to provide a true picture of the world given the constraints of the processing machinery of the brain. I suggest that the latter is more plausible. The system often has to operate under severe pressure and so may yield inaccurate results; nevertheless, the system as a whole is designed to operate veridically. Of

Table 2.7. *Design and results of the experiment by Shanks (1991a, Experiment 3)*

Condition	Trial types		Test symptom	Mean rating
Correlated	AB→O$_1$	(20)	A	32.3
	AC→no O	(20)		
Uncorrelated	DE→O$_2$	(10)	D	49.0
	DE→no O	(10)		
	DF→O$_2$	(10)		
	DF→no O	(10)		

A-F are the cues (symptoms) and O$_1$, O$_2$ are the outcomes (diseases); no O indicates no outcome. Numbers of each type of trial are shown in brackets.

course, when one understands the perceptual machinery at a fine enough level of detail, it should be possible to explain why illusions occur. In Chapter 4 we will see that many biases in associative learning can be understood in terms of the computations of a system which, fundamentally, behaves in a normative fashion.

In the remainder of this chapter I would like to discuss some further associative learning data that are rather harder to accommodate with the probabilistic contrast model and which indicate fairly clear biases in learning. We will try to see if there is any simple specification of the circumstances necessary to obtain biases such as these.

The first limitation of the probabilistic contrast model is that it fails in some circumstances to make any predictions at all because the task does not provide sufficient information about contrasts. For instance, in another medical diagnosis cue–outcome learning experiment that I conducted (Shanks, 1991a, Experiment 3), subjects were asked to rate the relationship between cue A and disease 1 and that between cue D and disease 2 after the training trials shown in Table 2.7. The procedure again required subjects to make diagnoses on each trial, with corrective feedback. For disease 1, the problem is easily mastered since the disease only occurs on AB trials; AC trials were accompanied by the absence of any disease. For disease 2 the task is ambiguous since both DE and DF were paired with the disease on 50% of trials during the learning task.

Subjects witnessed 80 trials on each of which one of the six trial types was selected at random. At the end of the learning phase, they were asked to rate the relationship between symptoms and diseases on a scale from 0 to 100. As expected, subjects gave a high rating of the B→O$_1$ relationship and a low rating of the C→O$_1$ relationship. E and F received intermediate ratings for O$_2$. The critical data are the ratings of A and D, and these are given in Table 2.7. Here, a significantly higher rating was given for the D→O$_2$

than for the A→O_1 relationship. A full explanation of this result must be postponed until Chapter 4, but for present purposes we need to see that this result is very problematic for the probabilistic contrast model. The reason is that the conditional probabilities needed by the ΔP rule are impossible to ascertain because there is no subset of events that allows the relevant computation to be made. Consider the A→O_1 relationship. The value of $P(O_1/A)$ can be calculated as 0.5 across the AB→O_1 and AC→no O trials, but since B and C are never presented alone, no suitable contrast probability is available. On the other hand, if the subject merely calculates ΔP across all experimental trials, then $P(O_1/A)$ is 0.5 and $P(O_1/-A)$ is 0.0, but this value is identical to that obtained for the relationship between cue D and outcome O_2.

In fact, it is not only the ratings for cues A and D that are problematic for the theory. When exposed to AB→O_1 and AC→no O trials, subjects had no difficulty deciding that B was strongly related to disease 1 and C was not. However, in terms of the probabilistic contrast model, it is not clear how these judgments could have been derived, since as with cue A, the relevant contrasts for B and C cannot be made. In order to compute the contingency between cue B and disease 1, the subject would have to have experienced some trials with cue A on its own, in order for the contrast with the AB trials to be made. No such trials were witnessed, though. In sum, there appear to be circumstances in which insufficient evidence is provided for the relevant contrasts to be made, but where subjects have no difficulty making associative judgments. Melz *et al.* (1993) have considered how the contrast model might be applied to such situations, but their suggestion is not very persuasive.

Another way in which biases may be observed is to manipulate the order in which trials of different types are presented. The statistic ΔP is based on values of $P(O/C)$ and $P(O/-C)$ calculated across a subset of trials in which everything is held constant except for the presence or absence of the target cue. Plainly, these probabilities are unaffected by the order in which the trials are presented, so long as the trial types themselves are the same. Thus on the normative theory, we would have to predict that the order in which the trial types are witnessed should make little difference to the observed judgments, since the probabilities are calculated across a set of trials and are unaffected by order.

Normative models are challenged by evidence suggesting that trial order *does* in some circumstances have an affect on associative learning. A case in point comes from a series of experiments reported by Chapman (1991) which we will have cause to return to in Chapter 4. One of Chapman's experiments will serve to illustrate the basic finding. In this study, subjects were exposed to a training procedure designed to establish one cue as having a negative contingency with the outcome. Using the standard medical diagnosis task, Chapman gave subjects 12 trials in the first stage on which

Table 2.8. *Design of the experiment by Chapman (1991, Experiment 4)*

Stage 1	Stage 2	Stage 3	Test trials
A→O	AB→no O		B
	CD→no O	C→O	D

A–D are the cues (symptoms), O is the outcome (disease), and no O indicates no disease.

symptom A was associated with a fictitious disease (see Table 2.8). In the second stage, 12 further patients had symptoms A and B but did not have the disease, and 12 patients had symptoms C and D and also did not have the disease. These AB and CD trial types were intermixed. Finally, in the third stage, 12 patients had both symptom C and the disease.

The A→O and AB→no O trials should establish cue B as having a negative relationship with the disease. According to the probabilistic contrast model, ΔP for cue B will be calculated just across the A and AB trials, yielding a value of -1.0, and in accordance with this prediction, cue B was given a negative mean rating of -42 on a scale from -100 to $+100$. Turning to the CD→no O and C→O trials, it is clear that apart from the order in which the trials are witnessed, the evidence presented to the subjects concerning the D→O relationship is exactly comparable to that concerning the B→O relationship, and that the value of ΔP is therefore the same for B and D.

Contrary to this prediction, Chapman found that subjects gave a significantly less negative rating for cue D (mean $= -34$) than they had for cue B, and she therefore concluded that trial order is an important factor in learning. Note that Chapman's experiment makes it unlikely that the effect is due to differential forgetting of the AB and CD trials, since the three-stage procedure ensured that these trial types occurred contemporaneously and hence should have been forgotten – if at all – to equal degrees. Clearly, subjects may be biased by trial order, but is there any obvious explanation of this effect? Consider a subject observing A→O trials followed by AB→no O trials. Given that cue A has been established as a predictor of the disease, the absence of the disease on the AB trials is surprising and should lead the subject to reason that the new cue, B, must have an effect that cancels out cue A, and hence must be negatively related to the disease. For the CD→no O, C→O trials, the subject has no prior expectations and so the absence of the disease on CD trials should be neither surprising nor unsurprising. Thus little should be learned on these trials, either positive or negative, about any contingency between symptom D and the disease. In Chapter 4 we will see that associationist theories posit a close relationship between surprise and learning, and even though in idealised circumstances such models can be

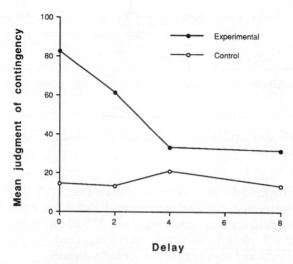

Figure 2.6. Mean judgments of contingency at different action-outcome delays. In the experimental conditions, P(O/A) was 0.75 and P(O/–A) was zero, and the outcome occurred after 0, 2, 4, or 8 s. In the control conditions, the sequence of outcomes in the corresponding experimental condition was played back to the subject independently of his or her responding. Subjects detected the action-outcome contingency at the 0 and 2 s delays but not at the 4 and 8 s delays. (After Shanks *et al.* 1989.)

said to compute ΔP, they can nevertheless explain Chapman's data on the basis that the difference in trial order alters the degree to which the outcome on CD trials, relative to AB trials, is surprising.

The effect of trial order seems to occur because the different trial types are presented in separate blocks. If they had all been intermixed, then of course trial order would have had no effect and judgments for the two target cues would have been identical since the trial types for B and D would have been functionally equivalent. Presenting trials in distinct blocks seems to allow biases in associative judgments to appear, and perhaps this is not surprising. If the learning mechanism needs prolonged exposure to a set of relationships in order to determine where the true predictive relationships lie, then this can only be achieved by continually repeating the various trial types in an intermixed fashion until learning is complete.

Returning to our main theme of biases in learning, it is important to note that contingency is not the only informational clue that the environment provides about associative relationships, and biases may emerge via the manipulation of other types of information. Temporal organisation is just such an alternative source of information. For instance, events close together in time are more likely to be related than ones separated in time. This contiguity factor was known by Hume in the eighteenth century but has been little investigated in the laboratory. In a simple experiment of ours

(Shanks, Pearson and Dickinson, 1989), judgments were dramatically reduced by the insertion of a delay between the action and outcome. In that experiment, which used a computer-based version of the instrumental procedure used by Wasserman and his colleagues, subjects pressed the space bar on a computer keyboard and judged the extent to which that action caused a triangle on the computer screen to light up for 0.1 s. There were four experimental conditions, lasting for 2 min, in which every action caused the outcome with probability 0.75. However, in different conditions it did so either immediately, or after 2, 4, or 8 s. As Figure 2.6 shows, when the outcome occurred immediately, the mean judgment of causality was 82.4 on a scale from 0 to 100. In conditions with delays of 2, 4, and 8 s the mean judgments were steadily reduced.

Plainly, increased delay caused a reduction in associative learning, but were the subjects still sensitive to the relationship at the longer delays? To answer this, we need to compare the judgments of subjects in the experimental conditions with judgments given under appropriate control conditions. But just what are the appropriate control conditions? The solution adopted in the Shanks *et al.* study was as follows. We recorded the pattern of outcomes which occurred in each experimental condition for each subject, and this was then played back to the subject in a later control condition independently of their responding. Thus the comparison between an experimental and control condition allows us to tell whether the subject was sensitive in the experimental condition to the causal relation at that action–outcome interval compared to the control condition in which there was no causal relation. Because the outcome occurred with the same frequency in the control and experimental conditions, this procedure allows us to unconfound the effects of a delay from the reduction in the number of outcomes that ensues if subjects reduce their response rate at longer delays.

Figure 2.6 shows that with a 4 s action–outcome delay, subjects were no longer sensitive to the instrumental associative relationship. Thus in this sort of task, people only appear to be able to detect a relationship between their actions and the outcome when the delay between them is less than about 4 s. It should be emphasised, of course, that much longer delays can certainly be tolerated in other situations. The slope of the contiguity function is likely to be highly task-specific.

Is it rational for subjects to reduce their judgments as the action–outcome delay increases? The first answer that comes to mind is 'yes'. It seems intuitively obvious that a delayed outcome is less likely to have been caused by a target event than an immediate one. However, it does not appear as if this intuition is captured by a supposedly normative analysis based on the statistic ΔP. In terms of contingency theory, delaying an outcome does not affect its probability. In each of the conditions of the Shanks *et al.* experiment, $P(O/A)$ was 0.75 regardless of the action–outcome interval. One solution is to take time into account when calculating $P(O/A)$ and $P(O/-A)$, in which case reducing the degree of contiguity between events will also reduce contingency. To see how this can occur, imagine dividing up one of the conditions in the experiment described above into equal time intervals, say of 1 s duration, and calculating the conditional probabilities with respect to those time intervals. When the action and outcome are highly contiguous, there will be a large number of time intervals in which both the action and outcome occur, so P(O/A) will be high and P(O/−A) will be low. As the action–outcome delay is increased, more time intervals will contain the action but not the outcome, and more will contain the outcome but not the action. The net result of both of these effects is to decrease ΔP.

But on closer inspection there appear to be two difficulties with this idea. First, remember that the theory must yield, for any given task, a specific value of P(O/A) and P(O/−A), and hence ΔP. But in order to specify what one means by the probability of an outcome given an action and the probability of the outcome in the absence of the action, it is necessary to specify across what window of time these probabilities are to be evaluated. Usually, the window is clearly demarcated in some way as a specific trial: for example, in the medical diagnosis procedure, each patient clearly constitutes a 'unit' as far as the computation of conditional probabilities is concerned. But in other situations, such as the free-operant task used in the Shanks *et al.* experiment, it is not. For such situations, there seems to be no independent or objective way of determining the time interval across which P(O/A) and P(O/−A) are to be evaluated. Yet the choice of this time interval makes a considerable difference to the predictions contingency theory makes. In any situation in which actions and outcomes are occurring across time, if we take longer and longer time windows then P(O/A) will approach 1.0, whereas if we take shorter and shorter time windows it will approach zero. Suppose that there is one cue, followed after 5 s by an outcome, and outcomes never occur noncontingently. What is ΔP? If we specify a time window of 10 s, then an action and its contingent outcome may fall inside one window and the contingency will be 1.0. But if the window is 1 s in duration, the action and outcome will fall inside different windows and the contingency will be negative: P(O/A) is zero, and P(O/−A) is less than zero, depending on how often the action is performed. Thus it is impossible to state definitively what the value of ΔP is, and this of course represents a major drawback to our normative theory.

The second problem is that even if a specific time interval for the calculation of P(O/A) and P(O/−A) can be specified, we would then have to predict that judgments should be discontinuous. Suppose, with respect to the Shanks *et al.* (1989) experiment, that the relevant time interval is 1 s. In the no delay condition, 75% of actions will fall inside a time interval that also contains an outcome, while no outcomes will fall inside a time interval that does not also contain an action. Accordingly, P(O/A) will be 0.75, P(O/−A) will be zero, and ΔP will be 0.75. In contrast, in the 2 s delay condition all actions will fall inside time intervals that do not contain an outcome, and all outcomes will fall inside intervals that do not contain an action. Hence ΔP will be zero. The conclusion is that judgments should be discontinuous and drop from 75 (on a scale from 0 to 100) to zero as soon as the action–outcome delay exceeds the time interval that has been specified. The data in Figure 2.6 do not support this prediction; instead, judgments fall steadily as the delay is increased.

Before leaving the topic of contiguity, it is worth mentioning that there is a small amount of evidence that spatial as well as temporal contiguity may influence learning. In one relevant experiment (Shanks, 1986), subjects were asked to judge the instrumental relationship between firing shells at a tank, and the tank's blowing up. On each trial, the tank passed through a gunsight on the right of the computer screen and the subject chose whether or not to fire at it. Subjects were told that because the tank was a long way off, it would not blow up immediately but after a short delay during which it would have moved some way across the screen. Subjects were also told that the tanks were traversing a minefield, which therefore represented an alternative cause of tank destruction. The probability of destruction given a hit was 0.75, while the probability in the absence of a hit was 0.25, yielding a contingency of 0.5. The experiment found that judgments of the action–outcome relationship were greater if the tanks blew up immediately adjacent to the gun-sight, rather than further away from it, even when temporal contiguity was held constant by varying the speed of the tanks. Thus there is some evidence of a role for spatial contiguity, but this is another factor that is not considered by current normative theories.

Contiguity is not the only way in which the temporal structure of events can vary. Another concerns the nature of the schedule that relates cues or actions and outcomes. An experiment by Reed (1993) highlights the effects of this factor. Reed was interested in the exact temporal patterning of outcomes that were contingent on an action. The task again involved judging the relationship between pressing the space bar on a computer keyboard and the illumination for 0.1 s of a triangle on the screen. In one condition, called the variable-interval (VI) condition, the computer first selected at random a time interval of between 1 and 20 s, and then arranged for the outcome to be contingent on the first response emitted after that time interval had elapsed. Responses occurring during the chosen time interval were ineffective. The number of responses emitted for each outcome was recorded and used to determine the schedule of outcomes in the second, variable-ratio (VR), schedule. Here, the first outcome occurred when the subject had made the same number of responses as had been emitted prior to the first outcome in the VI condition, the second outcome occurred when the same number of responses were made as had preceded the second outcome in the VI condition, and so on. Each condition lasted for 2 min and was followed by a rating of the action–outcome relationship on a scale from 0 to 100.

In this way, Reed was able to set up two conditions that were identical in terms of the overall action–outcome contingency but which differed in the precise temporal patterning of events. In both conditions, P(O/A) was 0.09 and P(O/–A) was zero, yielding a value of ΔP of 0.09. Thus if subjects were computing ΔP and basing their judgment on that statistic, the two conditions should have yielded equal ratings. In fact, judgments in the VI condition (mean 41) were significantly greater than those in the VR condition (mean 28).

Why should this have occurred? Reed (1992) reported two further experiments in which judgments differed in a pair of conditions again equated in terms of ΔP, and also obtained evidence that judgments depended on whether or not each outcome was preceded by a temporally-isolated response or not. Without changing ΔP, it is possible to have one condition in which the action immediately preceding each outcome is isolated from earlier actions, and a second in which the last action before the outcome is itself part of a dense sequence of actions. Reed found that the former case leads to higher judgments than the latter, and also showed that VI schedules tend to produce relatively large numbers of outcomes that are contingent upon temporally-isolated actions. It is important to acknowledge that the basis of VI/VR differences (which have been much studied in animals) is not entirely clear (see Dickinson, 1985), but whatever the explanation that will finally prove correct, it is obvious that such differences are at odds with models which rely on the computation of an overall metric of statistical contingency and which ignore the local characteristics of action–outcome pairings.

There are, in addition to contingency and contiguity, other factors that almost certainly influence associative learning but which have not been extensively studied in humans. For example, we have already seen, in the section on preparedness and phobias, that stimuli such as spiders and flowers seem to differ in the ease with which phobias are formed to them. One explanation (Seligman, 1971) is that over the course of evolution some stimuli have become 'prepared' in the sense that they are especially easy to associate with certain outcomes, particularly aversive ones. In fact, I argued that this result is not necessarily indicative of preparedness but could be explained in terms of socially- and culturally-transmitted beliefs that the probability of an aversive outcome is higher in the presence of such fear-relevant stimuli than it is in the presence of fear-irrelevant stimuli. But certain other examples of selective association are probably not explicable in these terms. Outside the field of phobias, it is known that cues can differ in terms of the ease with which they can be associated with a given outcome, the best-known example being taste-aversion learning. Animals readily associate gastric illness, induced by an injection of lithium chloride, with a novel food that they have eaten some hours earlier, but they find it difficult to learn an association between a tone and illness. In contrast, a tone will be far more readily associated with shock than will a novel food (Domjan and Wilson, 1972). There is clearly something in the nature of the stimuli themselves that operates in addition to contingency and contiguity. For example, the internal or external nature of a stimulus may be relevant: stimuli may be easy to associate when they are both internal (food and illness) or external (tone and shock), but not otherwise. As yet, such factors have not been widely studied in humans.

Summary

The environment provides us with a number of hints that events are causally or structurally related, amongst which contingency is the most obvious. Perhaps it is not surprising that humans have evolved to be quite finely sensitive (under ideal conditions) to this factor: after all, it is now known that associative learning in such lowly creatures as the mollusc *Hermissenda* is sensitive to variations in contingency (Farley, 1987). Nevertheless, the demonstration in humans of such sensitivity not only establishes the adaptiveness of the learning system, but also provides a fundamental empirical phenomenon against which theories of associative learning may be compared.

In this chapter we have seen that a good deal of associative learning data can be interpreted in terms of the application of quite simple statistical rules. Although data may be obtained that accord better with rules such as ΔD than with the normative ΔP rule, this may be attributed to pre-asymptotic learning. At asymptote, judgments tend to conform well to the ΔP theory. When additional cues are present, the ΔP theory must be interpreted within a framework such as Cheng and Holyoak's (1995) probabilistic contrast model. This specifies that ΔP should be calculated over a set of events in which except for the target cue or action, everything else is held constant. This account can accommodate a number of selectional results that have been taken as inconsistent with contingency theory, and to the extent that this is the case, associative learning can be interpreted as rational or normative.

Nevertheless, there plainly are situations in which biased judgments may be obtained. In situations where the conditional probabilities cannot be assessed across an unchanging background, but where subjects have no difficulty making associative judgments, the predictions of the account become unclear. When trials are presented in blocks rather than intermixed, effects of trial order can occur which are outside the scope of the theory, and biases can also be induced by varying the schedule relating the action to the outcome. Finally, I have argued that in some situations, particularly those involving manipulations of contiguity, the normative theory can come close to being undefined.

There is clearly a long way to go before we have a complete specification of what counts as a normative theory of learning, but it is already apparent that much of the data obtained in associative learning experiments can be viewed from a rational perspective. Our goal in later chapters, particularly Chapter 4, will be to see how apparent biases may emerge from a learning mechanism which has fundamentally evolved to yield normatively accurate beliefs concerning associative relationships. Our immediate concern, though, is to turn to the second of our basic questions about learning, concerning the nature of the representations that underlie associative learning.

3   Prototype abstraction and instance memorisation

In the last chapter we established that, to a first approximation, the human learning system behaves normatively. In attempting to answer the question 'What is the system doing?' (the first of our three questions), we have found that associative learning corresponds reasonably well to the prescriptions of contingency theories. In reaching this conclusion, we have remained agnostic about how the system actually works; all we have shown is that the behaviour it yields in associative learning tasks is roughly what a statistician utilising the notion of contingency would prescribe. In the present chapter we begin our consideration of how the system achieves this end. Here, we ask the second question, 'What sort of information is acquired during the course of a learning experience?'. In the next chapter, we will ask exactly how at the mechanistic level this information is acquired.

We begin by considering the phenomenon of generalisation, which represents one of the principal challenges to any theory of learning. Having learned something about one stimulus, how will acquired knowledge determine responding to some further stimulus? Generalisation is of interest not just because it is something we would like our theories of learning to explain, but also because it provides data that may tell us about the way in which information is represented. Two quite different views of the form of information underlying associative learning have been embodied in prototype and instance theories, and for these theories generalisation is a central issue. They attempt to describe how learning takes place in situations where there is considerable stimulus variation from trial to trial, and where the ability to generalise to new stimuli perceptibly different from ones already encountered is essential.

The concept of similarity

In order to understand how prototype and instance theories construe the learning process, and how they explain generalisation, it is necessary first to consider the concept of *similarity*. Essentially, prototype and instance theories assume that some mental representation is formed as a result of exposure to a set of training stimuli, with responding to further stimuli being a function of their similarity to the represented stimuli. Although similarity is a common everyday term, psychologists have developed a number of tools for measuring and analysing it; in particular, it has become commonplace to interpret similarity in terms of distance in a psychological space. As

Nosofsky (1992) has noted in a recent review, the idea of a psychological space and the development of accompanying techniques for analysing such spaces have proven to be amongst the most significant advances made in cognitive psychology in the last 40 years, since they allow us to discover regularities about that space that are distinct from regularities holding in physical space. This chapter reviews some of these developments.

How can we begin to investigate psychological spaces? As a simple illustration, suppose we have five stimuli that vary along two physical dimensions: say, rectangles varying in height and width. The assumption is that these stimuli are represented in the mind of an observer in a way that reflects his or her perceptual as well as cognitive capacities, with the representations of the stimuli not necessarily corresponding to their physical descriptions. For instance, a typical observer may be less able to make fine discriminations between rectangles varying in height than between ones varying in width, in which case two rectangles differing in height by 1 cm may seem more similar and may be mentally represented as closer together than ones differing in width by 1 cm. In order to discover how stimuli are represented in psychological space, we need to use a statistical method such as multidimensional scaling (MDS), which is one of a family of techniques for recovering the psychological structure inherent in a class of stimuli (Shepard, 1980). Subjects are invited to make pairwise similarity judgments concerning the stimuli; Figure 3.1 gives a hypothetical set of such judgments for the five rectangles.

The hypothesis is that these similarity ratings are monotonically-decreasing functions of distance in psychological space: the closer the points, the greater their judged similarity. Shepard (1958) showed that provided the judgments meet three constraints, they uniquely determine the relative spatial co-ordinates of the stimuli: that is to say, a given arrangement of stimuli can yield only a single set of similarity ratings and vice versa. The constraints are that the similarity s_{ij} between stimuli i and j be the same as that between j and i (symmetry; $s_{ij}=s_{ji}$); that for all stimuli $i, j, k, s_{ik} \geqslant s_{ij}+s_{jk}$ (triangle inequality); and that the similarity between each stimulus and itself is the same for all stimuli (minimality). There has been some debate about whether these constraints indeed hold for similarity judgments (see Nosofsky, 1992), but the success of the scaling approach suggests that violations of the constraints are the exception rather than the norm.

If we have a set of similarity ratings, it is then possible to recover the spatial locations of the stimuli using the procedure known as multidimensional scaling. Essentially, the procedure begins by assuming an arbitrary spatial arrangement of the stimuli and then determines a slight movement in the location of each stimulus so as to improve the overall correspondence between distance and similarity ratings. This procedure is iterated until no further improvement can be achieved. Eventually, a set of co-ordinates will be obtained such that the distance between each pair of stimuli correlates

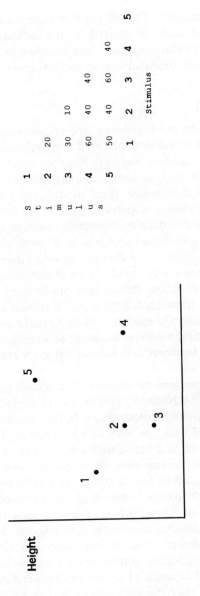

	1	2	3	4	5
1					
2	20				
3	30	10			
4	60	40	40		
5	50	40	40	60	40

Stimulus

Figure 3.1. Hypothetical stimuli and similarity ratings. The matrix on the right gives a set of hypothetical similarity judgments for pairs of stimuli (rectangles varying in height and width), assuming judgments are made on a scale from 0 (very similar) to 60 (very dissimilar). The figure on the left shows the locations of the five stimuli in psychological space which might be derived from a multidimensional scaling (MDS) analysis. As can be seen, the co-ordinates of the stimuli preserve the similarity structure shown in the matrix, such that the distance between a pair of stimuli approximately correlates with their judged similarity.

inversely with their judged similarity. The left panel of Figure 3.1 shows the hypothetical co-ordinates that might be derived for the rectangles. As can be seen, stimuli judged most similar appear close together in the spatial solution, and those judged most dissimilar appear furthest apart.

Identification learning

Similarity ratings have been the most common type of *proximity* measures for pairs of stimuli, but another measure that is of more relevance to associative learning involves identification learning. Suppose subjects are presented on each trial with one of, say, 12 stimuli and are required to learn to uniquely identify or name each stimulus. Thus, to stimulus 1 the subject must make response 1, to stimulus 2 response 2, and so on. After making an identification response to each stimulus, corrective feedback is provided if the response was incorrect, and stimuli continue to be presented until the subject has mastered the task. It is evident that just as with direct ratings of similarity, the probability of confusing stimuli in an identification learning task provides proximity data for those stimuli that can be used to recover the locations of each stimulus in psychological space. If stimuli 1 and 2 are very close in psychological space, the subject is likely to make many incorrect identification responses where response 2 is given to stimulus 1 and vice versa. In contrast, two stimuli far apart in psychological space are less likely to be confused.

As an illustration of this procedure, consider the data shown in Table 3.1 which are from an experiment by Nosofsky (1987). The 12 stimuli were discriminably-different reddish Munsell colour chips. In the experiment, each subject was presented with 324 trials, on each of which one of the stimuli, chosen at random, was presented and the subject was required to make the assigned identification response. In this case, the responses were the numbers 1 to 12, which were assigned to the 12 stimuli in a different way for each subject. Table 3.1 shows the complete matrix of confusions obtained in the experiment. Thus, on 665 occasions, subjects correctly gave response 1 to stimulus 1, but on 82 occasions they incorrectly gave response 3.

Using the procedure just described, Nosofsky reproduced the classic finding of Shepard (1958), that identification confusion data such as those given in Table 3.1 can be interpreted in terms of similarities between points in a multidimensional space. Specifically, Nosofsky performed an MDS analysis which yielded the points shown in Figure 3.2. The points shown in the figure are such that if the distances between all pairs of points are computed and rank ordered, then that rank ordering will very closely match the rank ordering of confusion frequencies. For instance, the second row of Table 3.1 shows that on trials with stimulus 2, response 4 was the most likely error and response 10 was the least likely. Accordingly, in Figure 3.2 the nearest neighbour to stimulus 2 is stimulus 4 and the furthest is stimulus 10.

Table 3.1. *Nosofsky's (1987) confusion data*

| | Response | | | | | | | | | | | |
Stimulus	1	2	3	4	5	6	7	8	9	10	11	12
1	665	17	82	13	63	19	4	20	13	14	4	7
2	21	670	38	121	9	20	12	10	7	1	3	6
3	152	28	453	37	82	78	7	39	18	10	7	5
4	12	156	35	581	19	46	36	12	14	2	1	4
5	73	8	63	12	552	32	13	100	24	18	13	9
6	30	15	85	42	55	466	46	70	59	12	15	19
7	10	17	10	38	12	54	616	17	117	4	9	15
8	14	9	35	20	77	64	18	513	28	89	33	14
9	6	5	26	13	16	65	101	48	507	10	52	69
10	5	7	8	7	14	16	9	50	8	767	22	5
11	3	3	8	3	7	6	11	37	28	48	594	172
12	4	5	10	0	7	2	18	22	27	13	216	591

Cell entries give the frequency with which each response was given to each stimulus. The stimuli were Munsell colour chips.

In MDS analyses such as this, it is possible to try spatial representations with any number of dimensions to see if the addition of extra dimensions improves the overall fit. Nosofsky was unable to get a better fit to the confusion data using a three-dimensional spatial representation of the stimuli than with the two-dimensional representation.

This sort of MDS analysis reveals a great deal about the psychological structure of these stimuli. For instance, the two dimensions recovered from the analysis happen to conform quite well to the brightness and saturation of the stimuli. According to the Munsell colour system, stimuli 5, 6, and 7 for instance are all equally bright, and this is reflected in their approximately equal values on the *y*-axis. Similarly, stimuli 2, 6, and 11 are all equally saturated, which is reflected in their approximately equal values on the *x*-axis. Thus the analysis suggests that colours are mentally represented in terms of these attributes. However, the analysis allows us to extend even

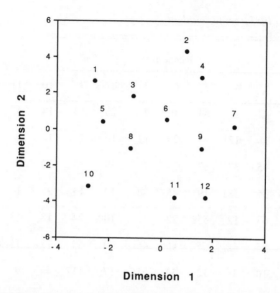

Figure 3.2. Locations in psychological space of the 12 stimuli in Nosofsky's (1987) experiment. The stimuli were Munsell colour patches and the co-ordinates were derived from a multidimensional scaling (MDS) analysis of the identification confusions shown in Table 3.1. The relative locations are such that the separation of two stimuli in this space predicts their likelihood of being confused.

further our understanding of identification learning. Multidimensional scaling only establishes that the probability of identification response j to stimulus i is a monotonic and decreasing function of the distance d_{ij} between them. What is the exact nature of this function? It has been established in a wealth of situations that provided the stimuli are readily discriminable, the relationship between similarity and distance is exponential (Shepard, 1987). If we assume that similarity and distance are related by the formula

$$s_{ij} = e^{-cd_{ij}} ,$$

$$(3.1)$$

where c is a free parameter, then we can predict each confusion almost exactly from the equation

$$P(R_j / i) = \frac{b_j \, s_{ij}}{\sum_k b_k s_{ik}} .$$

$$(3.2)$$

where b_j is a bias for response j. What this equation says is that the stimuli used in an experiment are represented as points in a n-dimensional psychological space, and the similarity between two stimuli is an exponential function of their distance. Then, the probability of giving response j to stimulus i depends on the similarity between stimuli i and j, divided by i's similarity to all of the stimuli. The denominator of this equation is simply a normalising factor that ensures that all of the response probabilities sum to one. Finally, it is assumed that the subject, other things being equal, may prefer some responses to others: this is reflected in the b_js.

To summarise: a set of identification confusions can be interpreted in terms of the pairwise distances between the representations of the stimuli in a psychological space. The perceptual system is assumed to analyse each stimulus and represent it in that space. The relative locations of the stimuli are then obtained by finding the set of points such that when the distance between any pair of stimuli i and j is computed and exponentiated, the likelihood of giving response j to stimulus i in a learning task is exactly or nearly exactly predicted. With this procedure, Nosofsky (1987) was able to account for an astonishing 99.8% of the variance in the data shown in Table 3.1. Plainly, the notions of spatial representation and exponential generalisation are very well supported.

Returning to our central interest, namely how identification responses are learned, Nosofsky's model as stated in Equation 3.2 assumes that the subject memorises the response that is associated with a given location in the space. As a result of receiving feedback concerning the correct response, that response is automatically attached to the corresponding stimulus location. When a stimulus is presented on a subsequent trial, its distance is determined from all the locations in the space that have responses attached to them, and the probability of emitting each of those responses is dictated by Equation 3.2. Note that if on an earlier trial the subject received feedback concerning the correct response for stimulus i, and on a subsequent trial that stimulus is again presented, then the distance is zero and the similarity is 1.0. However, this maximal similarity does not guarantee that the correct response will be made: if stimulus i is also similar to another stimulus j for which a different response is appropriate, then $P(R_j/i)$ may be less than 1.0 in Equation 3.2.

It should be clear from Equation 3.2 that the likelihood of making a correct response – that is, the probability of making the response that has been attached to a given location when a stimulus falling at that location is presented – depends ultimately on the parameter c in Equation 3.1. The similarity of a test stimulus to other stimuli that have different responses assigned to them depends on c, and as c increases, similarity decreases, in turn making correct responses more likely. Within Nosofsky's framework, then, learning is attributed to changes in similarity as well as to the simple encoding of instances. With more and more trials, the similarities of the stimuli to each other decrease and correct responses increase. This idea in fact has a venerable history going back to William James' (1890) elegant discussion of how repeated exposure to claret and burgundy, which are practically indistinguishable to the novice, tends to make them seem more distinct from one another such that to the expert wine-taster they are as distinct as chalk and cheese.

Given this reliance on the notion of changes in perceived similarity, it is natural to ask whether there is any independent evidence of such a process during identification learning. Unfortunately, there have been few successful attempts to show the effect directly, but a recent experiment by McLaren, Leevers and Mackintosh (1994) provides a fairly clear illustration. Subjects were trained in the first stage of the experiment to classify random-dot patterns into two categories. These patterns were in fact different distortions of two underlying prototype patterns, and all of the stimuli generated from one prototype were to be assigned to one category and those from the other prototype to a contrasting category. As training proceeded, subjects improved their classification accuracy. In the test stage, subjects now had to learn *identification* responses to each of two stimuli that were either selected from one of the categories seen in the first stage of the experiment or were novel; these pairs of patterns were objectively equally similar. McLaren *et al.* found that identification learning was faster for the previously seen stimuli than for the unseen ones, indicating that the seen stimuli had become more discriminable during the first stage of the experiment (so-called 'perceptual' learning), exactly in the way assumed by Nosofsky.

Prototype abstraction

At the outset of an identification learning experiment, the subject has had no experience with the stimulus–response assignments of the experiment and is thus forced to guess amongst the available responses. As a result of learning to identify each of a set of stimuli, the subject comes to have a representation of each stimulus and its associated response. When a given stimulus is presented, it is mentally compared with all stored stimulus representations and the most likely response is the one associated with the most similar stored representation. Of course, the assumption has been that each presentation of a given stimulus is effectively identical as far as the subject is concerned – each presentation of stimulus *i* corresponds to exactly the same point in psychological space. In real life, though, this situation is rarely, if ever, fulfilled. When learning a person's name, for instance, each observation of their face is quite different even though the face itself is objectively the same, and the different ways in which the face can present itself form a category rather than a single stimulus. We must therefore address the question of how such variability is dealt with. The more general case in which variability occurs is where several objectively distinct stimuli are associated with the same response or outcome, this being, of course, what we call 'categorisation'. So what happens when several different stimuli are associated with a single response?

One venerable approach is to say that categories of stimuli are represented by mental prototypes and that learning involves abstracting the appropriate prototype. The category *bird*, for example, might be repre-

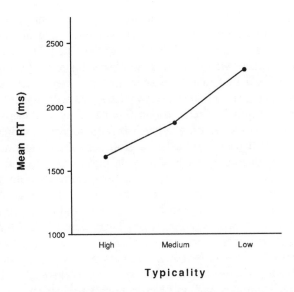

Figure 3.3. Mean reaction time (RT) in classifying stimuli as a function of their category typicality. Subjects were trained to classify dot patterns into different categories, with the patterns varying in terms of their level of distortion from the category prototypes. RTs for stimuli later judged highly typical of their categories were faster than those for stimuli later judged less typical. (After Rosch *et al.* 1976.)

sented by a typical bird that has been mentally abstracted from our experience of a large number of actual birds. On this account, responding to a new stimulus is a function of its similarity to the prototype. As test stimuli get closer to the prototype, they should therefore become easier to categorise, an effect that is readily demonstrated in the laboratory. For instance, Rosch, Simpson and Miller (1976) asked subjects to categorise artificial stimuli such as random dot patterns. A pattern from one of four categories was presented on each trial and the subject made a classification decision, with corrective feedback for incorrect responses. For each category, the patterns were the category prototype plus one pattern at each of five levels of distortion. After learning the category assignments, subjects were instructed to continue classifying the patterns as rapidly as possible and their response times were recorded. Finally, subjects rated each of the patterns in terms of how typical it was of its category.

On the basis that typical items are closer to the category prototype, the prototype view predicts that differences should be observable in responding to the stimuli as a function of their distance from the prototype, and this is exactly what Rosch *et al.* observed. As Figure 3.3 shows, items judged highly typical were classified more rapidly in the test stage than ones judged less typical.

A similar result is obtained in the classification of faces. Clearly, development equips us with the ability to discriminate between huge numbers of faces, even when they are highly similar. We might expect that such overwhelming exposure to faces would allow us to abstract a prototypical face, and indeed this is a view that has been advocated by researchers such as Valentine and Bruce (1986). On such a view, the speed with which a face that we encounter is classified as such should depend solely on its similarity to the face prototype, regardless of whether it is familiar or unfamiliar. To test this, Valentine and Bruce (1986) asked a group of subjects to rate on a scale from 1 to 7 how distinctive each of a set of unfamiliar faces was. On the basis of these ratings, faces were split into two groups containing the most typical and the most distinctive faces. Next, two different groups of subjects had to make classification decisions. On each trial either an intact or a jumbled face was shown, where jumbled faces were made from real faces whose features had been rearranged such that, for instance, the eyes were below the mouth. The subject had to press as fast as possible an 'intact' or a 'jumbled' response key. Half of the subjects were students for whom the faces were familiar, and half were students for whom they were unfamiliar. Thus, the same faces were used in the classification task for each group, but for half of the subjects they were familiar and for half they were unfamiliar.

The results, shown in Figure 3.4, confirm that faces closer to the prototype, i.e. typical faces, were classified more rapidly than distinctive ones, and this held both for familiar and for unfamiliar items. Valentine and Bruce also found that for the jumbled faces, it made no difference whether the face was made up from components of typical or distinctive faces.

Perhaps the most compelling reason to believe that abstraction of the prototype underlies categorisation is the abundant evidence that the prototype stimulus itself will be classified accurately and rapidly, even when it has never been presented in the training stage of an experiment. For instance, Homa, Sterling and Trepel (1981) trained subjects to classify geometrical patterns into three categories which varied in size. Three prototype patterns were defined, and training patterns were constructed by highly distorting these prototypes. In each block of the study phase, subjects saw 20 different patterns from category A, 10 from category B, and 5 from category C. One of these patterns was presented with corrective feedback on each trial, and subjects continued until they had achieved two errorless blocks. In the transfer phase, which occurred either immediately or after one week, the original training patterns plus the unseen prototypes were presented for classification.

Homa *et al.* found that subjects were in some cases more likely to correctly classify the prototype, which they had never seen, than any of the specific training instances, and this was particularly the case when the category contained a large number of instances (20). Figure 3.5 shows that the

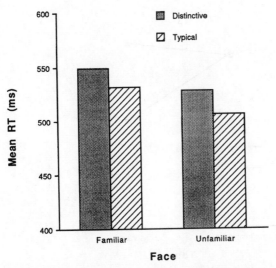

Figure 3.4. Mean reaction time (RT) to decide whether a stimulus was an intact or a jumbled face. Data are shown just for the intact faces, depending on whether they were familiar or unfamiliar and whether they were distinctive or typical. Regardless of whether the faces were familiar or not, subjects responded more rapidly to typical faces, that is, ones that were closer to the face prototype. (After Valentine and Bruce, 1986.)

benefit for the prototype over the original training items was enhanced when a long interval (one week) intervened between training and testing. Here, the prototype was correctly classified on 96% of trials, while the original patterns were only classified correctly on 85% of trials. Such results seem to imply that the prototype, at least in some instances, is mentally represented.

Homa *et al.* observed a further interesting result. When classification performance was tested after a delay of a week, considerable forgetting was evident for the original training items: for the 20-item category, performance fell by about 10%. In itself this result is not surprising, but as Figure 3.5 illustrates, Homa *et al.* found no such forgetting with regard to the prototype patterns. For these, performance if anything slightly improved across the delay. Thus prototype classification may continue to be highly accurate even when memory for the training instances has deteriorated, a result that is consistent with the notion that it is the abstracted prototype that is mediating classification.

A final finding in support of the idea of prototype extraction concerns the relation between recognition and classification. If all that is represented is the prototype, then subjects should be less likely to say that they recognise the original training items than the unseen prototype. To test this, Metcalfe and Fisher (1986) taught subjects to classify random-dot patterns into three

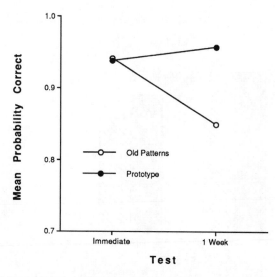

Figure 3.5. Mean probability of correct classification responses for original training patterns and novel prototypes. Subjects were trained to classify geometrical patterns into three categories. Either immediately or after one week, the training items and prototypes were presented as test stimuli. When tested after a week, subjects were more accurate in classifying the prototype than the original training stimuli. Also, the old training items were more susceptible to forgetting than the prototypes. (After Homa *et al.* 1981.)

categories. In the learning stage, each category contained six stimuli that were distorted from the relevant category prototype by equal amounts. At test, subjects had to classify and make recognition judgments for the old instances, the prototype, new instances that were distorted from the prototype by the same amount as the training items, and a large-distortion pattern.

Metcalfe and Fisher found that the prototype was again classified at least as well as the training items, but more importantly, the probability of falsely recognising the prototype was very high. In fact, it was even higher than the probability of (correctly) recognising the actual training items. Thus, consistent with the prototype view, subjects were apparently less likely to remember the training items than the prototype.

The results discussed so far in this section are all consistent with the idea that associative learning is mediated by prototype representations that are abstracted from the specific training stimuli, and such a view obviously has the attraction of great cognitive parsimony in that only a single representation needs to be maintained for each category of objects. But despite all of these findings, there are some well-known problems with the notion that concepts are represented by prototypes, so much so that it is now generally doubted whether prototype abstraction actually takes place in classification experiments. In the formation of a prototype, a large amount of informa-

Table 3.2. *Design and results of Medin et al.'s (1982) experiment*

	Pattern	Proportion category A responses
category A training items	1111	0.88
	1100	0.89
	0111	0.73
	1000	0.77
category B training items	0010	0.12
	0001	0.17
	1010	0.25
	0101	0.33
test items	0000	0.53
	0011	0.53
	0100	0.75
	1011	0.67
	1110	0.45
	1101	0.38
	0110	0.36
	1001	0.28

tion is discarded, yet this is information to which people can in fact be shown to be sensitive. For example, abstraction of the prototype means that information about the variance of the studied items, the number of such items, and correlations between values on different dimensions is lost. With regard to the last of these, consider the category of *birds*, for example. Within this category, there is a modal value on the size dimension, and a modal value on the singing dimension, and an abstracted prototype would encode these values. However, it is also the case that small birds are more likely to sing than large ones, a correlation between the dimensions that could not be recovered from the prototype. The evidence suggests that people are sensitive to such correlations, and therefore the notion that classification depends on distance to an abstracted prototype is at the very least inadequate.

As an illustration, consider an experiment by Medin *et al.* (1982). In the study phase subjects were presented with the four-dimensional stimuli shown in Table 3.2 which represented instances of two categories, where the dimensions encoded the presence or absence of each of four symptoms and the categories were fictitious diseases. The prototype of category A had a value of 1 on dimensions 1 and 2, while the prototype of category B had values of 0 on these dimensions. The remaining dimensions were not individually diagnostic, but instead had correlated values in the two categories.

Thus all category A exemplars had the same values on dimensions 3 and 4, while all category B exemplars had different values on these dimensions. By the end of the training stage, the category A items were all being assigned to category A with probabilities greater than 0.70, while the category B items were being assigned to category A with probabilities less that 0.35 (which means they were being assigned to category B).

The table shows that classification of test patterns was strongly affected by whether the values on dimensions 3 and 4 were the same or different. All test stimuli in which these values were the same were more likely to be assigned to category A, while all those in which the values differed tended to be assigned to category B, in accordance with the correlations of the study items. Pattern 1101, for instance, was classified in category B despite the fact that its values on the first two dimension were diagnostic of category A, while its values on dimension 3 and 4 taken in isolation were equally diagnostic (or nondiagnostic) of the two categories. Responding must therefore have been affected by the correlation between dimensions 3 and 4.

Although such results are problematic for prototype theories, it is of course possible that Medin *et al.*'s experimental task was not conducive to prototype abstraction and that under more favourable circumstances abstraction would occur. Certainly, given the results of the study of Homa *et al.* (1981), abstraction is more likely to take place in circumstances where a greater number of instances of each category is used than the relatively small number in Medin *et al.*'s experiment. However, an experiment by Ashby and Gott (1988) makes this line of reasoning look dubious since they obtained evidence of sensitivity to correlated dimensions even when a very large number of distorted patterns was used.

Ashby and Gott presented subjects with simple figures formed by a vertical and a horizontal line orthogonally joined at the upper left corner. Stimuli were distortions of two prototype patterns which were the same in terms of the length of the vertical line but differed in the length of the horizontal one. Subjects received 300 trials on which they had to assign each stimulus, with corrective feedback, into category A or B. As in Medin *et al.*'s experiment, there existed a correlation between the dimensions, and this is shown together with the prototypes in Figure 3.6. Stimuli varied at random about the prototype in terms of horizontal and vertical line length, but a positive correlation existed between these lengths. The ellipses in Figure 3.6 are 'iso-probability' contours representing stimuli that were generated from the prototype with equal probability. These contours indicate that for each category, greater horizontal line length tended to go with greater vertical length.

If classification is mediated by similarity to an abstracted prototype, then the dashed line in the figure should demarcate category A and category B responses. All points to the right of the dashed line are nearer the category B prototype, while all those to the left are nearer the category A prototype.

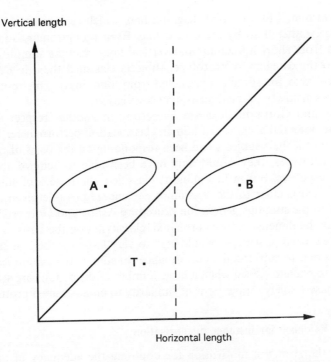

Figure 3.6. Schematic illustration of the stimuli used by Ashby and Gott (1988). The stimuli consisted of a vertical and a horizontal line joined at the upper left corner. On each trial, a distortion of one of the 2 category prototypes, A and B, was presented (with feedback) for classification. The distortions were generated by a procedure which maintained a correlation between the two dimensions, such that shorter horizontal length tended to go with shorter vertical length. The ellipses are 'iso-probability' contours representing stimuli that were equally likely to be generated. Category responses should be demarcated by the dashed line if subjects respond on the basis of similarity to the prototypes. In fact, the solid diagonal line divided subjects' category A and category B responses.

Thus stimulus T is nearer the category A prototype. In contrast, the solid diagonal line divides stimuli according to their likelihood of being generated from the two categories: points above the line correspond to stimuli more likely to have been generated from category A, and points below the line correspond to stimuli more likely to have been generated from category B. Because of the correlation between the values on the two dimensions, stimulus T is actually more likely to have been generated from category B than from category A.

According to the theory that classification is determined by similarity to abstracted prototypes, we should expect the dotted line to approximately divide the two response classes, even though there may be some noise and error in the subject's responding. The results Ashby and Gott (1988) obtained, however, show that this prediction is quite wrong: classifications were determined by the solid diagonal line, which represents optimal performance, rather than by the dotted line. Even though many stimuli constructed from short horizontal and vertical lines, such as stimulus T, were closer to the category A prototype, subjects classified them in category B (and vice versa for stimuli constructed from long lines). This result clearly indicates sensitivity to the dimensional correlation.

Ashby and Gott's findings are interesting in another respect that concerns the potential accuracy of human classification performance. They calculated that if their subjects had been responding on the basis of proximity to the prototypes, they would only have been able to achieve about 66% correct responses, but in fact subjects classified about 85% of stimuli correctly, which is close to the maximum percentage possible given the noise inherent in the stimulus-generation procedure Ashby and Gott used. Hence, whenever the dimensions are correlated, classifying on the basis of proximity to abstracted prototypes would lead to very large numbers of erroneous decisions and to sub-optimal performance. In sum, classification learning is difficult to explain – even when a large number of distortions are seen in the study phase – solely on the basis of similarity to an abstracted prototype.

Evidence for instance memorisation

Perhaps the strongest motivation for doubting the adequacy of prototype abstraction as the basis of category learning comes from experiments showing that information retained about specific training items influences classification. According to the logic of prototype theories, if the prototype is the only representation in memory that plays a role in the classification process, then specific training items that may have been studied should not affect classification performance. However, such 'exemplar' effects can be demonstrated in a number of ways. The simplest is to compare classification performance on old and new patterns equated for distance to the prototype. If classification is based purely on an abstracted prototype, then in experiments such as those of Rosch *et al.* (1976) and Homa *et al.* (1981), we should expect the original training items to be classified no better than new test items equidistant from the prototype. Instead, numerous experiments have found that old items are responded to better or faster than new items even when equated for similarity to the prototype.

Consider an experiment by Jacoby and Brooks (1984). They asked subjects to sort pictures of categories such as *cups* according to how typical they seemed of their respective categories. Subjects next classified as rapidly as possible 38 such objects shown in slides. Then in the test stage they classified new and old items, with some of the new items being the category prototypes. As Table 3.3 shows, prototypical items were classified more rapidly in the test stage than new different items. Moreover, they were also classified more rapidly than the original items had been in the study phase.

Table 3.3. *Results of Jacoby and Brooks' (1984) experiment*

Study phase		Test phase	
Study items	Prototypes	Study items	New different
268	237	214	278

Data are response times in ms.

But although it is beyond doubt that the prototype patterns were classified faster than the new different items, the fastest responses of all were to repeated study items. These were classified much faster in the test stage than they had been the first time round, which led to reaction times being faster than for the prototypes. If all the subjects were doing was comparing a test item to the prototype of its category, it is hard to see why the old study items rather than the prototypes should be responded to fastest.

Similarly, in Homa *et al.*'s (1981) experiment where subjects were trained to classify simple geometrical shapes into three categories of different sizes, the original patterns were classified better than new ones equidistant from the prototype. Subjects were tested on new patterns that were distorted from the prototype as much as the original patterns, but which systematically varied in terms of their similarity to specific old patterns. As Figure 3.7 shows, increased similarity to an old training item went with increased classification accuracy. If information about the specific training items is discarded in the formation of the prototype, it is difficult to see why such a result should emerge.

Direct evidence that training instances are retrieved in the process of classification comes from a series of studies by Malt (1989). She presented subjects in the first stage with pictures of artificial animals, varying on eight dimensions, for classification into two categories. Once these had been learned to criterion, a test stage was presented in which on some trials new items had to be classified. Malt reasoned that if the classification of a new item involved the explicit retrieval of an earlier similar training instance, then if that training instance was presented for classification on the immediately following trial, it should be classified – as a result of priming – more rapidly than it would be if it were dissimilar to the new item. This is exactly what Malt found: classification of old exemplars was faster if new but similar items had occurred on the preceding trial compared to the situation in which the animals were dissimilar. In addition, Malt obtained no priming in a second control condition. Here, subjects again classified each old instance in the test stage, but instead of classifying the new stimulus that occurred immediately beforehand, subjects simply had to judge whether it was large or small. This, of course, doesn't require accessing knowledge of old instances. The results from this condition also rule out the possibility that

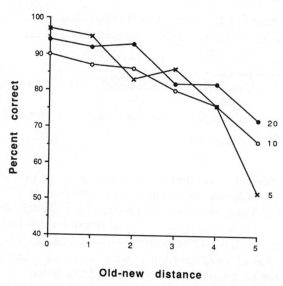

Old-new distance

Figure 3.7. Percentage of correct classifications for new stimuli varying in their distance from old training items. The experiment is the same one that generated the results shown in Figure 3.5. After learning to classify stimuli into one of 3 categories, which contained either 5, 10, or 20 stimuli, subjects were tested with new stimuli equated in their distance from the category prototypes but differing in their similarity to old training items. Stimuli similar to training items were classified much more accurately then those further away. (After Homa *et al*. 1981).

the new test stimulus speeded up processing of the following old instance simply as a result of sharing many attributes with it and hence causing perceptual enhancement. If perceptual enhancement caused the priming effects, priming should also have occurred in the condition where subjects make large/small decisions to the new animals, but it did not.

The evidence presented so far concerning the importance of instance memory in associative learning has come exclusively from categorisation experiments. However, instance effects can be demonstrated in other tasks too, and a particularly illuminating example comes from an ingenious study by Whittlesea (1987). He constructed the stimulus set shown in Table 3.4. A pseudoword (FURIG) was defined as the prototype, and various distortions around this prototype were constructed. Thus the type I words each differed from the prototype by one letter, the type II words by two letters, and the type III words by three letters. However, note that FURIG is the objective prototype for each set of words, in that it contains the modal letter in each position. For the Ia set, for instance, four of the five items begin with F, four have U as their second letter, and so on. Note also that while the IIa, b, and c items each differed from the prototype by two letters, the IIb items differ from the IIa items by one letter while the IIc items differ from the IIa items by two letters.

Table 3.4. *Stimuli used by Whittlesea (1987)*

Prototype	Ia	Ib	IIa	IIb	IIc	III
FURIG	FUKIG	FUTIG	FEKIG	FYKIG	FUKIP	PEKIG
	FUREG	FURYG	FUTEG	FUTYG	PUTIG	FYTEG
	PURIG	KURIG	PURYG	PUREG	FURYT	PURYT
	FYRIG	FERIG	FYRIP	FERIP	FYREG	FYKIP
	FURIT	FURIP	KURIT	PURIT	KERIG	KURET

The I stimuli differ from the prototype by one letter, the II items by two letters, and the III items by three letters. FURIG is the prototype of each letter set. Relative to the IIa items, the IIb items differ by one letter, the IIc items by two letters, and the III items by one letter.

Whittlesea did not require subjects to learn to classify these items, but instead used a speeded letter identification task. In the preliminary stage, the IIa, IIb, and IIc words were presented for 30 ms followed by a pattern mask which made identification very difficult, and subjects had to write down in their correct positions as many of the letters as they were able to read. This established a baseline against which later performance could be compared. In the study phase, the IIa words were presented for unlimited time and the subject merely had to write down each word. Then in the test phase the IIa, IIb, and IIc words were again presented for speeded classification as in the study phase.

Since the IIa items presented in the study phase are derived from the FURIG prototype, it is reasonable to imagine that subjects would abstract this prototype during exposure to the study items. But if that were the case, then the encoded prototype should facilitate the subsequent speeded identification of test words to the extent that they share letters with the prototype. Since the IIa, IIb, and IIc test words all differ from the prototype by two changed letters, they should therefore receive equal amounts of facilitation. Note also that the test sets are equated for the numbers of times specific letters appear in specific positions.

In contrast to this prediction, if the subjects have memorised the actual IIa study items and respond on the basis of similarity to those memorised exemplars then we would expect a greater degree of facilitation on the IIa test items than the IIb items, which in turn should show more facilitation than the IIc words. The IIa items should receive maximal facilitation, since they have themselves been memorised, the IIb items should receive less facilitation because they each differ from the IIa items by one letter, and the IIc items should receive least facilitation since they differ by two letters from the memorised exemplars. This result is exactly what Whittlesea observed. The top portion of Table 3.5 shows the degree of facilitation for each test type in

Table 3.5. *Results of Whittlesea's (1987) experiments*

Training stimuli	Transfer scores	
IIa	IIa	1.07
	IIb	0.80
	IIc	0.51
IIa	IIa	1.22
	IIc	0.65
	III	0.86

Numbers are the mean increase in the number of letters correctly identified, relative to the preliminary baseline stage.

this experiment. The figures in the table give the mean increase in numbers of letters correctly identified in position compared to the score from the preliminary phase. Since the maximum score on a test word is five letters correct in position, a facilitation of 0.5 represents a 10% increase in letter identification. The critical result was that there was a progressive decrease in facilitation from the IIa to the IIb to the IIc items, despite the fact that these were all the same distance from the FURIG prototype.

Whittlesea's stimulus set allows a further and even more damaging test of the notion of prototype abstraction. Suppose subjects are trained on the IIa items and then tested on the IIa, IIc, and III items at test. The new aspect is the inclusion of the type III test items, and for these, we would have to predict poorer performance than for the other items if similarity to the prototype underlies facilitation, since the type III items differ from the prototype by three letters while the IIa and IIc items differ by only two. The results of this second experiment, also shown in Table 3.5, show the direct opposite of this in that the type III items receive significantly *more* facilitation than the type IIc items. Why should this be? Inspection of the stimuli reveals that each of the type III items is constructed so as to differ from a IIa study item by at most one letter, while the IIc items differ by two letters. If facilitation is a function of the similarity of a test item to a memorised study item, then this is exactly the outcome that would be expected. Whittlesea's results therefore not only provide a further illustration of instance memorisation, but also demonstrate that increasing the proximity of an item to the prototype can have a detrimental effect if the item becomes less similar to the study items, as in the comparison between the IIc and III items.

Given these results, it is perhaps not surprising that researchers have begun to acquire evidence of instance effects in language tasks as well. For example, Jared, McRae and Seidenberg (1990) compared the time required to name written words such as CAVE and CANE. They found slower response times for the former than for the latter and interpreted this in terms of the similarity of other known words. While both words have regular spelling–sound correspondences, CAVE is similar to a word (HAVE) which has a different pronunciation, whereas all _ANE words are pronounced like CANE. The difference in naming latency can therefore be understood in terms of a model in which words are represented as points in a multidimensional space, with performance being affected by other similar words having the same or a different pronunciation.

The context model

On the basis of the overwhelming evidence that instance memorisation plays a role in category learning, Medin and Schaffer (1978) proposed that a significant component of the mental representation of a category is simply a set of stored exemplars or instances. The mental representation of a category such as *bird* includes representations of the specific instances belonging to that category, each presumably connected to the label *bird*. In a concept learning experiment, the training instances are encoded along with their category assignment.

Medin and Schaffer assumed that both instance storage and prototype abstraction could occur during the learning of a concept, and of course any combination of these two processes is possible. However, subsequent studies have shown that performance in a great many category learning studies can be understood in terms of instance storage alone and that the notion of prototype abstraction may be unnecessary. For example, it is not necessary to cite prototype abstraction in order to explain the accurate and rapid classification of prototype stimuli: as originally noted by Hintzman and Ludlam (1980), instance theories can account for prototype effects, because as a test item gets closer to where the prototype would be, its summed similarity to the training instances also increases. Thus classification decisions based on summed similarity will be maximal for the prototype pattern. If instance storage is known to take place, and if instance storage models are able to explain a broad range of empirical phenomena, then adding a prototype abstraction process adds little in terms of explanatory power while reducing the parsimony of the theory.

Although there remain advocates of prototype extraction (e.g. Homa, Dunbar and Nohre, 1991) as well as sceptics about the distinguishability of prototype and instance theories (e.g. Barsalou, 1990), the evidence seems to go against the notion that category learning is based on prototype abstraction, and instead supports instance memorisation. The instance view proposes that subjects encode the actual instances during training and base their classifications on the similarity between a test item and stored instances. When a test item is presented, it is as if a chorus of stored instances shout out how similar they are to the test item. At the formal level, the best-developed such theory is the *context model* of Medin and Schaffer (1978) and Nosofsky (1986), which is directly related to the theory of identification learning we considered earlier. The basic idea is simple: the probability of assigning a stimulus *i* to category *J* is a function of the summed similarity of *i* to each of the members of category *J* that have been stored in memory, divided by the summed similarity of stimulus *i* to all exemplars of all *K* categories:

$$P(J/i) = \frac{\sum\limits_{j} s_{ij}}{\sum\limits_{K} \sum\limits_{k} s_{ik}} \tag{3.3}$$

In this equation, s_{ij} is the similarity of stimulus *i* to stimulus *j*, *j* being one of the stored exemplars of category *J*. The name 'context model' denotes the fact that an instance is a complex conjunction of the target item together with the current context.

We can illustrate the power of instance-storage theories by considering the results of a category learning study by Nosofsky (1987) that utilised the identification data described earlier. Recall that Nosofsky was able to obtain the psychological co-ordinates (see Figure 3.2) of each of a set of 12 colour patches by use of a multidimensional scaling procedure. With other subjects, Nosofsky then examined category learning. These subjects were trained to classify the stimuli across 240 trials into two categories with appropriate feedback, and the relevant data are from the last 120 trials. Three of the classifications are shown in Figure 3.8. In the pink–brown problem (so-called because the members of the one category are shades of pink and those of the other category shades of brown) the members of the two categories are discriminable by a boundary going from top-left to bottom-right. In the diagonal classification, four stimuli on the diagonal are in category 2 and the rest in category 1. Note that this classification cannot be solved by a single linear boundary. Finally, in the brightness problem the members can be classified on the basis of a horizontal boundary. In this latter problem, stimuli 1, 3, and 5 were presented without explicit feedback and therefore represent transfer stimuli.

Since Nosofsky knew the psychological locations of each stimulus, he was therefore able to compare subjects' classification decisions with the predictions of the context theory. For each stimulus, its distance to each other stimulus was computed and these distances were exponentiated as specified in Equation 3.1. Then, the probability of assigning the stimulus to category 1 was determined by Equation 3.3. Nosofsky found a remarkable degree of concordance between predicted and observed classifications, with over 99%, 97%, and 99% of the variance in the observed classifications in the pink–brown, diagonal, and brightness classifications being accounted for, respectively. Figure 3.9 shows the observed and predicted probabilities of making a category 1 response for each stimulus, including the transfer

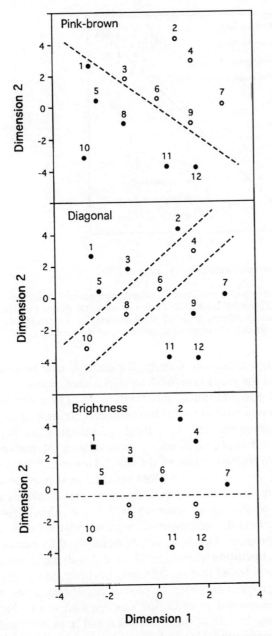

Figure 3.8. Three classification problems used by Nosofsky (1987). In each case, the same set of 12 stimuli (Munsell colour patches) was used. The locations of these stimuli in psychological space come from a separate MDS analysis (see Figure 3.2). In each problem, subjects were trained to classify some stimuli (filled circles) into category 1 and others (open circles) into category 2. In the brightness problem, filled squares represent transfer stimuli. Note that the diagonal classification cannot be solved by a single linear boundary.

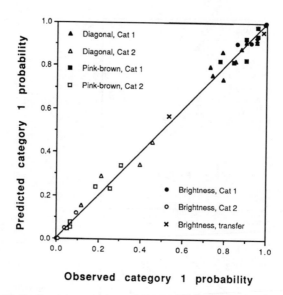

Figure 3.9. Predicted (context model) and observed classification probabilities for the classification problems shown in Figure 3.8. Each symbol refers to one of the stimuli from one of the problems. Note that in general category 1 stimuli (filled symbols) were correctly assigned to category 1 and category 2 stimuli (open symbols) to category 2. (After Nosofsky, 1987.)

stimuli in the brightness condition. Clearly, the concordance between the subjects' responses and the predictions of the model is exceptional.

Nosofsky obtained one additional important result. The psychological co-ordinates of the stimuli were obtained from subjects performing an identification learning experiment, and from these co-ordinates can be computed the distances, and hence classification probabilities, of the subjects who performed the categorisation part of the study. However, except in the pink–brown problem, these distances did not on their own provide a very good fit to the data. Instead, Nosofsky had to assume that for the categorisation subjects, the psychological space in which the stimuli fell was stretched and shrunk along its component dimensions relative to the space for the identification subjects. That is to say, Nosofsky had to assume that selective attention was operating when subjects were classifying the stimuli.

To illustrate, Nosofsky found that no stretching of the space was required for the pink–brown problem in order to predict classification responses, whereas a considerable amount of distortion was required in the diagonal and brightness problems: in both of these, the space had to be stretched vertically and shrunk horizontally. This makes good sense. In the brightness condition, especially, it is clear that subjects can solve the classification by ignoring the saturation dimension (dimension 1) and attending instead to the brightness dimension (dimension 2). Variations in saturation are irrelevant for the purposes of classification, while variations in brightness, especially in

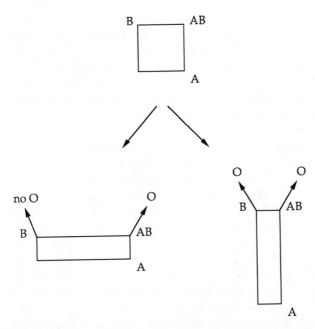

Figure 3.10. Explanation of contingency effects provided by instance theories. Two independent cues, A and B, are represented as points in a psychological space where the end-points of each dimension are the presence and absence of the cues (top panel). When the outcome (O) occurs given the compound cue AB but not (no O) given B alone (lower left), the space is stretched along the relevant A dimension and shrunk along the irrelevant B dimension, making A more similar to AB. In contrast, when both AB and B are accompanied by the outcome (lower right), A becomes less similar to AB.

the vicinity of the category boundary, are highly significant. By ignoring the saturation dimension, subjects are effectively shrinking the space horizontally such that stimuli differing only in saturation become more similar. Conversely, stretching the space vertically makes stimuli differing in brightness appear more distinct and hence less confusable.

It is clear that subjects could solve some of these classifications by abstracting a prototype, and so to test this, Nosofsky compared the fits of the context model with those of a prototype theory, assuming that the training instances formed the basis for an abstracted prototype. For the pink–brown condition, 95% of the variance was accounted for, and although this is quite good, it is statistically much worse than the fit of the context model. For the diagonal problem, on the other hand, only 33% of the variance was accounted for. Clearly, the formation of a single category 1 prototype would not allow that problem to be solved. In fact, the fit in this condition was still poor (82%), relative to that of the context model, when it was assumed that two category 1 prototypes were abstracted, one for stimuli 1, 2, 3, and 5, the other for stimuli 7, 9, 11, and 12.

Having found such an impressive degree of support for the instance approach, it is now appropriate to return to our fundamental effect – the effect of contingency on associative learning – and ask how this account explains such effects at the informational level. The basic idea is illustrated in Figure 3.10. The figure shows stimuli as points in a psychological space, with the dimensions of the space corresponding to the dimensions of variation amongst the stimuli. Since we are interested in the simple case in which we have two stimuli, A and B, we assume that each stimulus is represented by an independent dimension on which the values are absent and present. In a design in which two types of trials (AB and B) are presented, the relevant stimuli occupy the left (B) and right (AB) upper vertices of the space.

The explanation of contingency effects is straightforward. Suppose AB is paired with the outcome while B on its own does not predict the outcome. According to the instance account, attention will selectively be directed to the A dimension, since it is presence versus absence of this cue that predicts the outcome. In terms of the psychological space depicted in Figure 3.10, the net effect of this is to stretch the space horizontally and shrink it vertically, such that more weight is given to the A dimension. In contrast, when AB and B trials are both accompanied by the outcome, the A dimension is irrelevant and it is now the B dimension to which attention should be directed. Hence the space is stretched in the vertical direction and shrunk in the horizontal direction.

The outcome is that presentation of A on its own will elicit a greater response tendency in the case where the outcome is contingent upon A than in the case where it is not contingent upon A, since, as the figure shows, A is more similar to AB for the situation shown in the lower left-hand panel than for that shown in the lower right-hand panel. Since AB is associated with the outcome, this greater degree of similarity translates into a greater response tendency. Hence we have a simple explanation at the informational level of why the contingency effects described in the last chapter emerge.

Categorisation and recognition

It has occasionally been claimed that data from experiments comparing recognition and categorisation pose a problem for the view that categorisation is based on comparison to memorised instances. We have seen how memorised instances are assumed to determine classification, but what about recognition? Here, the idea would simply be that an item on a recognition memory test is called 'old' to the extent that it is similar to one or more of the memorised instances, regardless of their category assignment. On the face of it, we might expect to see some relationship between categorisation and recognition, because on the instance view the same information underlies both sorts of decision. For instance, we might expect subjects to be more accurate at classifying items they believe to be old, since such items are more likely to have been stored in memory and hence be capable of influencing categorisation than items the subject thinks are new. In other words, we might expect the probability of correctly classifying an item given that it is believed to be old, P(correct/'old'), to be greater than the corresponding probability for items believed to be new, P(correct/'new').

Following such logic, Metcalfe and Fisher (1986) tested the instance approach by comparing P(correct/'old') and P(correct/'new'). One aspect of Metcalfe and Fisher's study was discussed previously. Remember that in the training phase they presented six random dot patterns generated from three category prototypes, and then tested subjects with the original training patterns, the previously-unseen prototypes, and other novel patterns. For each test pattern, the subject first had to decide whether it was old or new, and then assigned it to one of the categories. We saw earlier that subjects classified the prototype very accurately and that they were very likely to falsely recognise the prototype. But the interesting finding for our present purposes concerns the comparison of P(correct/'old') and P(correct/'new'). Since subjects had to make both classification and recognition decisions, it is possible to calculate the probability that a subject makes a correct classification decision, given that he or she has called the item 'old', and compare this with the corresponding probability for items called 'new'. Overall, the difference between P(correct/'old') and P(correct/'new') in Metcalfe and Fisher's experiment was 0.08, reflecting little dependence between classification and recognition. Thus, in contrast to the apparent prediction of instance-based theories, subjects were barely more likely to correctly classify items they recognised than ones they did not.

Moreover, it has even been observed that recognition can be virtually at chance in situations where categorisation is excellent. In an experiment in which the stimuli were schematic faces varying on four dimensions, Medin and Smith (1981) trained subjects on nine study items, five from category A and four from category B. Subjects made a classification decision for each face before receiving corrective feedback, and were trained either until they had reached a criterion of one errorless run through the nine items, or until they had received 32 complete runs through the stimulus set. In a test phase, the nine study items plus seven new items were presented for recognition and classification. As a result of the extended practice, Medin and Smith's subjects reached an overall probability of correct classifications in excess of 0.70 in the test phase. However, recognition was extremely poor. The probability of correctly saying 'old' to an old face, 0.82, only just exceeded the probability of mistakenly saying 'old' to a new face, 0.77.

Are these results genuinely at variance with instance theories? Although Metcalfe and Fisher's logic seems at first glance to be sound, it turns out that instance theories can predict exactly the results obtained by Metcalfe and Fisher and by Medin and Smith. Let us suppose, following the memory model of Gillund and Shiffrin (1984), that recognition is a function of the summed similarity of a test item to all memorised items. Imagine that a test item is equally and highly similar to two memorised instances from different categories. Because of the high degree of similarity, the item will be called old, but it may at the same time be classified at chance because of its equal similarity to instances from competing categories. In contrast, a test item may be only moderately close to instances from a certain category, and hence called new, while at the same time being highly dissimilar to instances from other categories: hence, it may be classified with some accuracy. Such intuitions show that instance theories do not necessarily predict a relationship between recognition and categorisation.

More formally, Nosofsky (1988) performed a simulation of Metcalfe and Fisher's experiment in which hypothetical stimuli were generated from three category prototypes. In the first stage instances from the categories were memorised, and in the second stage old and new items were presented for recognition and classification. Nosofsky obtained a difference between P(correct/'old') and P(correct/'new') close to zero, thus reproducing Metcalfe and Fisher's results. The intuitive notion that instance theories predict a relationship between classification and recognition turns out to be incorrect.

Artificial grammar learning

The instance memorisation view of associative learning can encompass a truly impressive amount of experimental data, often at an exquisite degree of quantitative detail. While much of the data comes from experiments using very simple stimuli such as Munsell colour patches (as in Nosofsky's study), I shall try to illustrate the power of the instance approach by considering recent examinations of artificial grammar learning, a task that was discussed briefly in Chapter 1.

Artificial grammar learning simulates language learning (albeit in a highly simplified way) by using relatively complex stimuli, but more importantly is usually thought to require something altogether more sophisticated than memorisation of exemplars. Thus it would be very persuasive if we could show that instance memorisation plays a significant role in this task. Recall that in a typical experiment, during the study phase subjects read strings of letters (e.g. VXVRXR) generated from a grammar such as that shown in Figure 1.1. The task is to memorise the strings. Then prior to the test phase, subjects are told that the strings were generated according to a complex set of rules and that their task in the test phase is to decide which new items are also constructed according to those rules. Subjects then make grammaticality decisions for grammatical and nongrammatical test items. The typical result is that subjects are able to perform above chance on such a grammaticality test (the typical level is about 60–70% correct classifications). On the notion that learning consists of abstracting the underlying

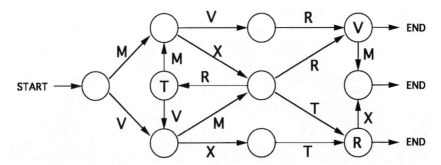

Figure 3.11. The artificial grammar used by Brooks and Vokey (1991) to generate letter strings. The grammar is entered at the left-hand side and links are traversed until the grammar is exited on the right-hand side. Each link yields a new letter which is added to the string. If a node contains a letter, then the letter can be added repeatedly at that point in the string. Strings such as MXRVXT can be generated from the grammar.

composition rules of the grammar, transfer of this sort should indeed be expected, for the same reason that a listener can judge novel sentences of a natural language grammatical. It should be obvious why artificial grammar experiments have been taken as highly simplified models of the learning of natural languages.

As was mentioned in the Introduction, much of the interest in artificial grammar learning has been driven by the question of whether it is possible to learn unconsciously about the rules of a grammar (see Reber, 1989; Shanks and St. John, 1994), but that is an issue that I shall not address here. Instead, I focus on the equally interesting question of what exactly is learned. Do subjects learn abstract grammatical rules or do they store the training strings as instances? Reber's (1989) view has been that abstract rule learning is the principal determinant of classification performance, but recent examinations have revealed that much of the data from artificial grammar learning experiments can be explained on the assumption that subjects merely store the training strings as whole items in memory, and respond to test strings on the basis of similarity to the encoded study instances.

As an illustration, Brooks and Vokey (1991) trained subjects on strings generated from a grammar and tested them on new strings, half of which were also from the grammar and half of which were nongrammatical in that they could not be generated from the grammar. They divided these test strings into those which were similar to study items (differing by only one letter) and those which were dissimilar (differing by more than one letter). A sample of the strings is shown in Table 3.6 and the grammar is illustrated in Figure 3.11. Thus consider the training string MXRVXT. The test string MXRMXT differs by one letter while the string MXRTVMR differs by four letters, yet both are grammatical. The test string MXRRXT differs by

Table 3.6. *A sample of Brooks and Vokey's (1991) strings*

Study items	Similar test items		Dissimilar test items	
	Grammatical	Nongrammatical	Grammatical	Nongrammatical
MXRVXT	MXRMXT	MXRRXT	MXRTVMR	MXRTVMM
VMTRRRR	VMTRRRX	VMTRRRT	VXVTRRR	VXVTRRM
MXTRRR	VXTRRR	TXTRRR	VXTRRRX	TXTRRRX
VXVRMXT	VXVRVXT	VXVRTXT	VXVR	VXVM
VXVRVM	VXVRVV	VXVRVT	VMRVMR	VMRVMM

The study items and grammatical test items can all be generated from the grammar shown in Figure 3.11.

one letter and the string MXRTVMM differs by four letters, yet both are nongrammatical.

The key finding was that while subjects were able to perform quite well overall on the grammaticality task, making 60% correct decisions, Brooks and Vokey found a large effect of similarity on grammaticality decisions. Indeed, while most (64%) similar grammatical strings were called 'grammatical', only a minority (45%) of dissimilar grammatical strings were classified as grammatical. This result is consistent with the idea that training strings are memorised during the learning phase, with new strings being classified on the basis of similarity to stored strings, but the result is at variance with the claim that classifications are based simply on abstracted grammatical rules: similarity does not objectively affect grammaticality. It is true that grammatical status appeared to have an effect on grammaticality judgments that was independent of similarity, since more grammatical than nongrammatical close items were called 'grammatical' (64% versus 42%) and more grammatical than nongrammatical dissimilar items were called 'grammatical' (45% versus 28%). But even this result can be questioned. As we will see in the next chapter, with a more sophisticated measure of similarity, Perruchet (1994) was able to eliminate any evidence of an independent contribution of grammaticality over and above similarity.

One final point is worth making concerning the nature of the entity that, according to instance theories, is memorised. Thus far I have simply assumed that study items like MXRVXT are represented in some form that allows overlap with other items such as MXRMVRV to be computed, but it is clear that this can be achieved in many different ways. For instance, since the study strings are presented visually, perhaps they are memorised in some visual code. Or, given that subjects probably articulate the letters in the string, they may be stored in an articulatory or phonological code. Yet

Table 3.7. *Results of Whittlesea and Dorken's (1993) experiment*

Test operation	Response	
	Grammatical	Nongrammatical
Same	0.66	0.34
Different	0.49	0.51

Data are the response probabilities for grammatical test items.

again, they may be encoded in some amodal manner. Each of these forms of representation would allow a different measure of similarity between pairs of items to be determined, since strings differ visually, phonologically, and on many other dimensions. Thus we might expect the memorial representation to be a complex entity that codes many different aspects of the study strings.

In fact, there is evidence that the representation encodes not only the various physical aspects of the study strings, such as whether they were presented visually or auditorily, but also the types of mental operations that were performed upon them during the study phase. This is illustrated in an elegant experiment by Whittlesea and Dorken (1993). Subjects studied items such as ENROLID that were generated from a grammar, prior to making grammaticality decisions for novel test strings such as ONRIGOB, which was in fact grammatical, and OGALPAD, which was not. The interesting aspect of the experiment was that subjects had to treat different items in different ways. Half of the study items had to be pronounced aloud while half had to be spelled. At test, each item was again either spoken or spelled prior to being judged grammatical or nongrammatical.

As a result of items being spoken at study and spoken at test, or spelled at study and spelled at test, the subjects' ability to determine grammatical status was very good. Table 3.7 shows that when the study and test operations were identical, the majority of grammatical test strings were called 'grammatical'. In contrast, a mismatch between study and test operations led to subjects being incapable of making accurate grammaticality decisions.

These results suggest that what is encoded in memory is a highly specific entity that retains all sorts of detailed information about the item and about how it was treated at study. From the point of view of associative learning, the inference is that when an item is presented which contains structural relationships between a set of experimenter-defined elements, what is stored in memory is a 'snapshot' that not only preserves those structural relationships, but also many of the relationships pertaining at that moment between the elements of the stimulus and the mental operations performed upon them, as well as perhaps such things as the experimental context and emotional state of the subject. Hence as a model of instance-based learning, the context model is aptly named.

Forgetting

On the face of it, the idea that all the instances of a category that a subject observes are encoded in memory and retrieved in the process of making a category decision is, to say the least, counter-intuitive. Certainly, with the exception of very unusual people such as Luria's (1968) famous mnemonist S., who could apparently remember everything that had ever happened to him given sufficient retrieval clues, we are not normally aware of the multitudinous exemplars we have encountered in the past and are, in fact, often unable to explicitly recall them. Obviously, the matching processes envisaged by instance theories must be assumed to take place unconsciously, but even with this proviso, it seems farfetched to imagine that separate memory traces of encountered events are preserved.

However, the memorisation view receives support from another finding which may at first sight seem even less plausible. If one adopts an instance approach, then it would appear that forgetting represents something of a problem, if by forgetting we mean the genuine loss of information from memory. Surely we forget most of the exemplars of a category we encounter? Since we are assuming that all previously-encountered items are available when a classification decision has to be made, how can we at the same time deal with forgetting, which requires the disappearance of some memory traces with the passage of time? The answer is that there is good evidence that forgetting – meaning the genuine loss of information – *hardly ever occurs.*

It is important to note, as a preliminary point, that the term 'forgetting' is ambiguous and has at least two distinct meanings. First, we use the term descriptively to refer to a person's behaviour. Of course, at the behavioural level, forgetting definitely does occur, if by this we mean a subject's inability to retrieve items of previously-learned information. Secondly, the term is used in an explanatory sense, to mean that information has been lost or erased from memory, and we usually attribute forgetting (in the behavioural sense) to forgetting (in the explanatory sense). What we are concerned with here is whether forgetting in the latter sense does in fact occur.

Traditionally, it has been assumed that all forgetting (as behaviourally defined) from long-term memory is caused by *retroactive interference* (RI). Suppose a subject has learned an A–B association, and then learns a new and contradictory association between A and C. Retroactive interference refers to the fact that the subject's ability to remember the original A–B relationship is impaired as a result of having learned the new association, because in some way or other the A–C association has interfered with the A–B association. Such an interference account is to be contrasted with a pure decay process, whereby information is lost merely as a result of the passage of time, a phenomenon for which there is little supporting evidence. As we will see below, subjects are more likely to forget an A–B association when it is followed by an A–C than by a control C–D item, despite the fact that the time interval between study and test is equated. This finding cannot be explained on the notion of trace decay because the A–B association should have decayed equally whether it is followed by an A–C or by a C–D trial.

Typically, the process of interference (and hence forgetting itself) has been attributed to one of two sources. The first is that the A–C association may have 'overwritten' the earlier A–B association and hence led to unlearning, with the original association literally lost or at least fragmented as a result of the later learning. The second possibility, more congenial to instance memorisation theories, is retrieval failure: the interfering information may have made the original association difficult to retrieve, without it actually having been lost. Experimental evidence suggests that the extent to which information is genuinely lost or unlearned is negligible, and instead, forgetting is almost entirely attributable to retrieval failure.

As support for this claim, consider the results of some experiments conducted by Chandler (1993) which used the A–B, A–C design with the elements of the associations being forenames and surnames. In Chandler's first experiment, subjects initially read a series of A–B target names (e.g., Robert Harris) and were given an immediate cued recall test (Robert H–?) to ensure that they had learned them. Some of the targets were experimental items and some were control items: this refers to the fact that on a second list of names, which also had to be learned, there appeared A–C names (e.g., Robert Knight) that were similar to the earlier experimental names, but there were none on the second list that were similar to the earlier control names: instead, there were a number of unrelated C–D names. Then, on a final memory test, subjects were given a mixed-up list of the first names and surnames that had appeared in the original list (the A and B elements), and were asked to match them up. Note that on this test, the potentially-interfering surnames (e.g. Knight) from the second list did not appear. Chandler found that subjects were able to correctly match 59% of the control names but only 46% of experimental names, representing a sizeable RI effect.

We obviously have evidence of interference-induced forgetting in this experiment, in that subjects were poorer at remembering names when they were followed by other similar names, but what is the basis of this RI effect? On an unlearning account, the effect is attributed to the fact that the similar names (Robert Knight) led to the earlier names being unlearned or

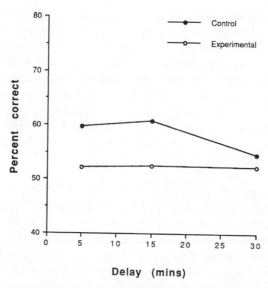

Figure 3.12. Percentage of correct responses at various delays for control and experimental items. Subjects were presented with A–B pairs which were followed by C-D (control) or A-C (experimental) pairs. At test, subjects had to match the original A and B items. At short delays, performance for control items was much better than for experimental items, suggesting that retroactive interference (RI) from the A-C items prevented retrieval of the experimental A-B pairs. At longer delays, however, the RI effect was absent. This implies that the RI seen at shorter delays was not due to unlearning of the experimental A-B pairs but was instead due to retrieval failure. (After Chandler, 1993.)

overwritten in memory, with the A–B association being permanently lost. In contrast, the retrieval failure account attributes the effect to the fact that the names from the second list have somehow blocked retrieval of the target names.

How can we discriminate these hypotheses? In a further experiment using the same general procedure, Chandler obtained a result that cannot be accounted for if the target words were really lost from memory. In this study, Chandler merely varied the delay interval between the two lists of names and the final memory test for different groups of subjects. In one case, the delay between the first list and the test was 5 min (during which the second list was learned), in a second condition the delay was 15 min, and in a final condition it was 30 min. Once again, experimental names on the first list were followed on the second list by similar names while the control names were not. For the groups tested after 15 or 30 min, the second list was presented immediately after the first list and was then followed by some filler tasks before the test took place.

The results are shown in Figure 3.12, which reveals a surprising finding, namely that the amount of RI actually declined as the retention interval was increased. Retroactive interference is apparent whenever performance is worse for the experimental than for the control items. At intervals of 5 and 15 min, the results of the first experiment were replicated, with RI of about 8% being obtained. However, by 30 min there was no evidence whatsoever of RI. Such a result is impossible to explain on an unlearning account, because the account has to attribute the RI obtained at the shorter retention intervals to genuine unlearning of the original experimental items, at least relative to the control items; but if the experimental items have to some extent been unlearned, they should still be harder to recall than the control items at the 30 min delay.

In sum, Chandler's results, which have also been obtained with pictorial stimuli (Chandler, 1991), suggest that forgetting is not in the main due to real unlearning or fragmentation of memory traces but is instead due to the fact that later information blocks the retrieval of earlier information. Although it is somewhat tangential to our current concerns, Chandler's results suggest that this blocking process only occurs when the potentially-interfering information is active in memory. As the delay before the test is increased, the interfering information gets less active in memory and is less likely to block retrieval of the target trace. In a last experiment, this interpretation was confirmed. Chandler (1993) presented the final memory test 30 min after the original A–B list, but for different groups gave the second list either just after the original study list or just before the test. The outcome was that no RI occurred in the former case – replicating the result from the second experiment – but that RI did occur when the interfering list was learned just prior to the test. Such a result confirms the idea that blocking of the target items by subsequent items only occurs when the latter are active in memory. When the intervening items occur long before the test, sufficient time has elapsed for them to become inactive in memory and hence unable to block retrieval of target names.

To the extent that instance memorisation theories require access to large numbers of stored instances, the results of the present section are congenial. It appears that the behavioural phenomenon of forgetting is not due to a genuine loss of stored memory traces. Hence it is not as implausible as it might seem to suppose that people have implicit access to stored instances.

Representation and learning in instance theories

I suggested in the Introduction that an adequate theory of learning needs to state what type of information is learned as well as what sort of mechanism governs the learning process. The general thrust of the findings reviewed in this chapter is that learning in many contexts can be interpreted as the simple memorisation of relatively unanalysed training instances, with responding being based on computed similarity to those stored items. The evidence certainly argues that the representational claims of instance theories allow a great deal of data to be explained, but I end this chapter by discussing some

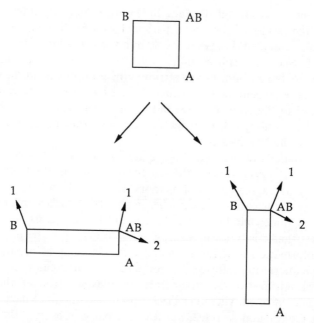

Figure 3.13. Schematic illustration of Gluck and Bower's (1988) experiment. Cue compound AB is paired with outcomes 1 and 2 on equal numbers of trials, while B is paired with outcome 1. Possible ways in which the space might be distorted as a result of selective attention are shown in the lower left and right panels. Note that A is always at least as similar to the instances of category 1 as to those of category 2.

evidence suggesting that pure instance memorisation may not be entirely sufficient as a description of the way information is encoded in associative learning tasks. To pre-empt somewhat the course of the discussion, in the next chapter we will see that a specific type of learning mechanism (in fact, an associationist one) can account for some of these discrepancies while still at the same time basically memorising the training instances.

Recent data have suggested that it is not sufficient to assume that each instance is automatically stored and that new stimuli are simply categorised on the basis of similarity to those stored exemplars. One piece of evidence comes from influential experiments originally conducted by Gluck and Bower (1988), and subsequently replicated and extended by Nosofsky, Kruschke and McKinley (1992), and Shanks (1990). There has been some controversy concerning the design originally adopted by Gluck and Bower (see Shanks, 1990, and Gluck and Bower, 1990), but the critical result is quite well-established. Before describing one of the actual experiments, I illustrate the basic effect. The design is shown in Figure 3.13.

The figure shows a psychological space corresponding to the presence and absence of two stimuli, A and B, with the top right-hand corner corresponding to the stimulus combination AB. Suppose subjects are presented with a series of AB trials, on half of which AB is associated with outcome or category 1, and on half of which with outcome 2. Suppose also that on a number of additional trials, B is associated with category 1. The question is, what will the subject do on a test trial with cue A? Instance theories, and to some degree common sense, predict that subjects will either be indifferent between categories 1 and 2 for this cue, or will associate it with category 1. The reason is straightforward: there are an equal number of AB→1 and AB→2 traces, so A is equally similar to each subset of those traces. There are also some B→1 traces, and if A is at all similar to B, then this will tip the balance towards category 1 as the choice response, but if there is no similarity between A and B, then A will be assigned to categories 1 and 2 with equal probability. Selective attention to the dimensions will clearly not alter these conclusions.

What happens in an actual experiment? I shall describe the results of a study of my own (Shanks, 1990). Tests of this prediction have been somewhat complex, using the simulated medical diagnosis procedure described earlier. Subjects made one of two diagnoses on each of 160 trials for patients having between zero and four different symptoms such as swollen glands. Because there are four possible symptoms, there are a total of 16 different patterns of symptoms. One disease occurred three times as often as the other, and these are referred to as the 'common' and 'rare' diseases respectively. Corrective feedback was given on each trial. The common disease is said to have a higher 'base-rate' than the rare disease.

In order to determine which symptoms would be present on each trial, the computer controlling the experiment began each trial by selecting either the common or the rare disease to be the correct disease on that trial, with the probability of selecting the common disease being 0.75 (see Table 3.8). Then a set of symptoms was selected according to a complex set of probabilities. One target symptom corresponds to cue A and was paired with the common and rare diseases on an equal number of trials. On many trials, this symptom occurred in conjunction with other symptoms. If the common disease was selected for a given trial, then the target symptom was chosen with probability 0.20, while if the rare disease was selected, it was chosen with probability 0.60. The third pair of rows in Table 3.8 show the converse of these figures, namely the probabilities of each disease given the target symptom.

The crucial point of the design is that the probability of the common and rare diseases were equal on trials with the target symptom. The equality P(common disease/target symptom)=P(rare disease/target symptom)=0.5 held whether all trials with the target symptom were considered or just those consisting of the target symptom on its own. Thus, objectively, on a target symptom trial the subjects should have predicted the rare and common diseases with equal probability. However, in the absence of the symp-

Table 3.8. *Design and results of the study by Shanks (1990)*

P(common disease)	0.75
P(rare disease)	0.25
P(target symptom/common disease)	0.20
P(target symptom/rare disease)	0.60
P(common disease/target symptom alone)	0.50
P(rare disease/target symptom alone)	0.50
Proportion common disease diagnoses	0.37
Proportion rare disease diagnoses	0.63

The first three pairs of rows give programmed values. The final pair gives subjects' mean probability of choosing the common and rare diseases across the last four trials on which the target symptom alone occurred.

tom, the common disease occurred more frequently than the rare disease. These trials, on which other symptoms were present, correspond to the B→1 trials from Figure 3.13.

The results across the last four trials on which the target symptom occurred on its own are shown in Table 3.8, which gives the observed probability with which the rare disease was chosen. The critical result is that subjects were more likely (p=0.63) to predict the rare than the common disease on trials with the target symptom, contradicting the prediction of instance theories. Thus contrary to our instance-based analysis of the situation shown in Figure 3.13, subjects are more likely to choose outcome 2 rather than outcome 1 on a test trial with stimulus A, and Nosofsky *et al.* (1992) have confirmed at the quantitative level that the context model is unable to predict this outcome.

The reason why the effect occurs is presumably a reflection of a well-known phenomenon in animal learning that we shall explore much further in the next chapter, namely that cue A is a better predictor of category 2 than of category 1. Because of the B→1 trials, cue B is a very good predictor of category 1, and hence on the AB→1 trials A fails to be associated with category 1. At the same time, the occurrence of category 2 on some of the AB trials contradicts the expectation that cue B predicts category 1: therefore, A is a useful predictive event (for category 2) on AB trials.

A similar result which again raises questions about the informational assumption of instance theories comes from studies by Medin and Edelson (1988). Again using the medical diagnosis task, they presented subjects with AB→1 trials and AC→2 trials, where A, B, and C are symptoms and 1 and

2 are fictitious diseases. As before, the subject's task was to make a diagnosis on each trial, with corrective feedback being given. Like Gluck and Bower (1988), Medin and Edelson were interested in whether subjects would be sensitive to base-rate information about the diseases, and accordingly, they presented subjects with three times as many AB→1 as AC→2 trials. Suppose that after mastering this classification, subjects are now asked to make a diagnosis for a patient who has just symptom A. Instance theories, and again common sense, suggest that disease 1 is more probable. The context theory makes this prediction because, other things being equal, A is equally similar to the memorised AB and AC instances, but since there are more of the former, the overall degree of similarity is greater for the disease 1 than for the disease 2 instances. This is exactly what Medin and Edelson found: subjects tended to diagnose disease 1 on an A test trial, and did the same on an ABC test trial too. Again, the conjunction ABC is more similar to the instances of category 1 than to those of category 2.

The critical result, though, came from test trials on which patients had symptoms B and C, which had never occurred together in the learning phase. The prediction of the context model is again that subjects should choose disease 1, because again BC should be more similar to the AB than to the AC instances, but in fact the subjects were significantly more likely to choose disease 2. Such a result is quite at variance with the context model and other instance theories because there should be no way in which the study trials have made BC more similar to the AC than to the AB traces. Why does the effect occur? One possible explanation appeals to the fact that during training symptom A is likely to become more strongly associated with disease 1 than with disease 2. This in turn might allow symptom A to compete more effectively with symptom B as the cause of disease 1 than with symptom C as the cause of disease 2. Thus C would be able to maintain a stronger association with disease 2 than symptom B does with disease 1, which then explains behaviour on the BC test trial. Evidence supporting this explanation comes from a study (Shanks, 1992) confirming that symptom A is indeed more strongly associated with disease 1 than disease 2 after the training trials.

Gluck and Bower (1988) and Medin and Edelson (1988) have made a very important contribution to our understanding of associative learning by introducing experimental designs that allow the informational assumptions of instance theories to be examined. The upshot is that despite all the successes of instance theories, these assumptions appear to be questionable.

Generic knowledge

A final piece of evidence that is problematic for instance theories comes from a very different source. One of the best-known distinctions in cognitive psychology is between so-called 'semantic' and 'episodic' memories, a distinction first drawn by Tulving (1972). Episodic memories essentially correspond to what I have been calling 'instances'. They are snapshots of experience that are stored in a relatively unanalysed form. Semantic (or generic) knowledge, on the other hand, represents facts about the world and about language and is not tied to particular episodes. For example, knowing that Paris is the capital of France is a piece of semantic knowledge.

A very considerable amount of effort has been put into trying to determine whether episodic and semantic memories are psychologically distinct, and there is no doubt that there are some good reasons for believing this to be the case. For example, many neuropsychologists have argued that the sort of amnesia suffered by Greg (Chapter 1) is a selective impairment of the ability to retrieve episodic memories (see Shallice, 1988, ch. 15). Without getting into the details of this controversial area, it is important to recognise that semantic knowledge presents something of a problem for instance-based theories. Such theories would either have to argue that so-called semantic memories are really just collections of episodic memories, or would have to concede the existence of a separate form of knowledge over and above stored instances. But neither of these options is particularly palatable. To concede the existence of a separate form of knowledge is to admit that the theory we have discussed at such length in this chapter is radically incomplete, while to argue that so-called semantic memories are really just collections of episodic memories flies in the face of some very compelling evidence to the contrary.

Here, we will consider just one piece of evidence that suggests that semantic knowledge cannot be understood in terms of stored instances. Watkins and Kerkar (1985) obtained evidence from an ingenious set of experiments that when an item is presented twice, recollection cannot be explained by reference to the retrieval of two separate memorised instances. In their studies, subjects were presented with a list of words such as 'umbrella', with some words being repeated on the list. In one of the experiments, words were written in different colours such that each of the once-presented words and each occurrence of a twice-presented word was in a different colour. At the end of the list subjects were first given a brief distracting task and then required to remember as many of the words as possible.

As would be expected, and as the left side of Figure 3.14 shows, Watkins and Kerkar observed better recall of twice-presented than of once-presented words, but they also observed a phenomenon called *superadditivity*. Let us assume that the probability of remembering *either* presentation of a twice-presented word is just the probability of remembering a once-presented word. If the probability, P2, of remembering a twice-presented word is just the probability of remembering each separate occurrence, then it should be readily predictable from the probability, P1, of remembering once-presented words:

$$P2 = P1 + P1 - P1.P1. \tag{3.4}$$

That is, the probability of remembering a twice-presented word is just the probability of remembering its first presentation, plus that of remembering its second presentation, minus their product. The latter is subtracted because if the subject remembers both presentations, that will not increase recall probability compared to a situation in which they only remember one presentation. Watkins and Kerkar found that P2 was actually greater than predicted by this equation (i.e. superadditivity occurred). The probability of recalling a twice-presented word, P2, was much greater (0.46) than the value predicted (0.32) from Equation 3.4 given the value of P1, which was 0.18. This indicates that the assumption that P2 is straightforwardly related to P1 is incorrect: the probability of remembering, say, the first presentation of a twice-presented word is greater than that of remembering a word only presented once, which in turn means that the second presentation makes the first more recallable, and vice versa.

This result may not, on its own, be too problematic for instance theories. Although they provide no mechanism whereby memorised items may affect each other's recallability, it is perfectly possible that such a process may exist and that subjects in Watkins and Kerkar's experiment were retrieving stored instances. But a further aspect of the results makes this look extremely unlikely. If each stored trace of a twice-presented item is enhanced in its memorability compared to that of a once-presented item, then we would expect this to extend to *all* aspects of the trace: specifically, subjects should be better able to remember the colour a twice-presented word appeared in on its first or second presentation than that of a once-presented word. To test this, after subjects had attempted to recall the words, Watkins and Kerkar gave them a complete list of the words with the twice-presented words marked, and asked subjects to say what colour (or colours in the case of the marked words) each had appeared in. As Figure 3.14 shows, Watkins and Kerkar observed exactly the opposite of the predicted result. Subjects were significantly *poorer* at remembering the colours associated with twice-presented words than once-presented ones. After a correction for guessing, the probability of recalling the colour of a once presented word was 0.37, while the probability of recalling one of the colours associated with a twice-presented word was 0.25.

Watkins and Kerkar interpreted this result in terms of the formation of generic or semantic memory traces. Subjects are assumed to abstract from multiple presentations of a stimulus only those aspects that are invariant, and discard trial-by-trial fluctuations. Such an abstraction occurs even with as few as two presentations. While generic memories like this are beyond the scope of instance-based theories, it turns out that they are very readily

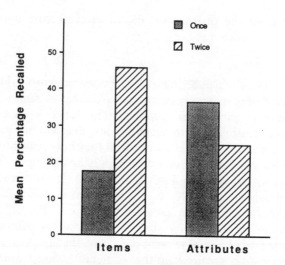

Figure 3.14. Mean percentage of items and attributes recalled for once-presented and twice-presented words. Subjects studied lists of words on which some were repeated and others were not, with each word being in a different colour. At recall, subjects remembered the twice-presented words ('items') much better than the once-presented ones. However, they remembered the colours ('attributes') of the twice-presented words more poorly than of once-presented words. This result is at variance with the idea that performance was based on the recollection of stored instances. (After Watkins and Kerkar, 1985).

explained by the sort of associationist learning mechanisms that we consider in the next chapter. Such mechanisms deny that semantic and episodic memories are stored in distinct systems, but nor do they treat semantic memories as collections of episodic ones. Rather, semantic knowledge emerges from episodic information in a more complex and constructive way.

Summary

In this chapter we have seen that the technique of multidimensional scaling provides an extremely powerful tool for analysing identification and categorisation data, because it formalises the notion that stimuli can be represented as points in a multidimensional space whose dimensions are the psychologically-significant dimensions of variation amongst the stimuli. Within this sort of spatial representation, it is possible to formulate prototype and instance theories as answers to the question 'What is learned?'. Although certain phenomena such as the accurate classification of an unseen prototype are congenial to prototype theories, other phenomena are more problematic. In particular, the abundant evidence that memorised instances play a role in categorisation is inconsistent with the idea that prototype extraction is the sole basis of concept learning. While there may be circumstances in which such a process does occur, there seems to be little evidence that cannot instead be more parsimoniously understood on the basis of instance memorisation alone.

Instance or exemplar theories such as the context model are capable of accounting for a wealth of empirical phenomena, including enhanced responding to the unseen prototype, and the successes of such theories strongly encourage the view that instance memorisation plays a role in category learning. Nevertheless, there is evidence questioning the notions that (i) responding to a stimulus is a function of similarity to previous instances and that (ii) all aspects or elements of an instance are equally likely to be memorised. In the next chapter we will turn to our third question, 'What is the mechanism of learning?'. We will see that some of the ambiguities concerning instance memorisation dissolve when we look in more detail at the learning mechanism. We will examine possible mechanisms of associative learning that preserve the essential spirit of instance memorisation as the basis of learning, but which are also capable of explaining the fact that some aspects of a stimulus may be better preserved than others.

4   Connectionism and learning

In the Introduction I argued that human associative learning can best be understood by considering three questions. What does the system do? How, in broad informational terms, does it do it? And how is this achieved at the mechanistic level? We have answered the first question by suggesting that to a first approximation the system does what a statistician would do: that is to say, it computes the degree of conditional contingency between events as defined by the metric ΔP. The second question is answered (again to a first approximation) by reference to the memorisation of instances, with stimuli being represented in a multidimensional psychological space and with inter-stimulus similarities being an exponential function of distance in the space. We now turn to our final question: what is the mechanism of learning? What sort of mechanism computes contingency, and represents associative knowledge in terms of memorised instances? In this chapter we will examine how contemporary associationist learning systems attempt to answer this final question.

I also pointed out in the Introduction that enthusiasm for associationist accounts of human learning waned somewhat in the 1960s and 1970s. Partly, this was due to impatience with the highly constrained tasks that researchers in the verbal learning tradition employed, but partly it was also due to the apparent inability of associationist theories to cope with more complex examples of human learning and cognition. For instance, a number of extremely sophisticated analyses had been undertaken of the difference between novices and experts performing various skills such as playing chess or remembering strings of digits (e.g. Chase and Ericsson, 1981; Chase and Simon, 1973). Experimental studies indicated that the sorts of things experts learned that allowed them to outperform novices were the ability to recognise particular configurations of pieces, the ability to organise knowledge into a complex hierarchy, and the ability to rapidly search a decision tree. There seemed to be simply no way in which the primitive processes of associationist theories of learning could illuminate such skills. Also, what interest there was in associationism was dealt a very severe blow by the publication in 1969 of Minsky and Papert's book *Perceptrons*, which demonstrated to the satisfaction of many that associationist devices were incapable in principle of learning a number of elementary functions that are well within the scope of human learning. I shall return to this later.

However, during these decades, when more complex symbolic theories were being developed and explored, associationism was alive and well in the animal learning laboratory, and was also influencing researchers like Rosenblatt (1962), Anderson (1968), and Willshaw, Buneman and Longuet-Higgins (1969) who were interested in how the human brain manages to store large amounts of information. These researchers noted that the brain is essentially an associationist system in that its basic computing element, the neuron, simply transmits electrical excitation or inhibition from its dendrites to its axon. Clearly, networks of highly interconnected neurons do manage to store large numbers of separate memories. But how is this achieved?

The answer that has emerged from this expanding field of research is that information is stored in a distributed fashion in weighted connections between neurons. At the psychological level, this viewpoint proposes that associative learning is represented in the form of mental associations between the elements of cues, actions, and outcomes, and that these associations are incremented or decremented on a trial-by-trial basis according to a 'connectionist' or 'adaptive' learning rule. It might seem that this approach offers nothing beyond the sorts of associative theories that were present 30 or 40 years ago, and therefore – as Minsky and Papert (1969) pointed out – would suffer from the same limitations, but two important steps have been taken. One is that the specific rules determining the amount by which an association is changed at a particular moment are vastly more sophisticated than earlier ones, and the other is that contemporary connectionist models are able to learn internal representations that boost to a great extent their learning capabilities. Such internal representations will be considered later in the chapter.

Connectionist models containing very many highly interconnected units have a number of general characteristics that immediately commend them. First, the only ways in which information is transmitted in an adaptive network are via the excitatory or inhibitory influence of one unit on another, and of course these are precisely the processes by which neural activity is propagated in the brain. Plainly, it is an attractive feature of any psychological model that it appeals only to processes known to operate in the brain.

Secondly, connectionist networks are by their nature parallel processing devices, in that the connectivity of individual units allows parallel sources of information from other units simultaneously to influence the state of activation of any given unit. The appeal of such parallel processing comes from the following observation. People are extremely good at retrieving stored knowledge from partial cues, even when some of the cues are inappropriate. For instance, a friend can often be recognised even if part of their face is occluded or if they are wearing an unfamiliar pair of glasses, a word can be made out even if poorly enunciated, and so on. This kind of memory is called *content-addressable*, because part of the content of the memory is used as the cue for retrieving the whole item, and it is particularly easy to achieve in parallel systems (and correspondingly difficult to achieve in conventional non-parallel ones).

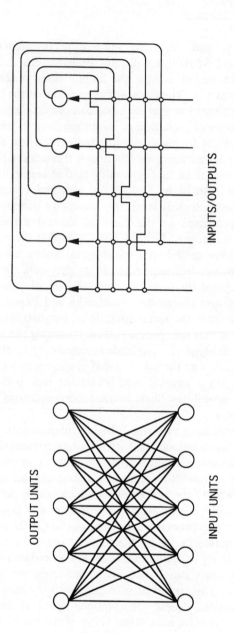

Figure 4.1. Left: A feedforward network. A homogeneous layer of input units is connected in parallel to a layer of output units. The connections between the units have modifiable weights. When a pattern of activation (representing the input stimulus) is applied to the input units, activation spreads to the output units via the weighted links between the units. A learning algorithm adjusts the weights on a trial-by-trial basis until the correct pattern of activation is obtained on the output units. Right: an autoassociator. Here, there is only a single layer of units dealing with both input and output. Each unit provides an input to every other unit but not to itself.

OUTPUT UNITS

INPUT UNITS

INPUTS/OUTPUTS

Imagine that memories are stored as patterns of activation across a large number of units. When at a later point some subset of those units is reactivated in parallel to represent the retrieval situation, the connections between the units will allow excitation and inhibition to spread such that the entire original pattern of activation is recreated. Even if some units are inappropriately activated, the original pattern may still be recreated if it represents the best 'solution' to the current set of input cues. Essentially, connections can be seen as constraints that exist on the spread of excitation and inhibition through the system.

A third attractive feature of adaptive connectionist networks is that knowledge is distributed across very many connections. Each connection represents a relationship between a pair of what we might best call *micro-features*, and hence it requires a large number of connections to represent memory for a complex object such as a face. At the same time, any given connection can contribute to many different memories. In short, a network can store a large number of superimposed patterns with each one being distributed across the network and with each connection contributing in a small way to very many memories. As a consequence, networks tend to demonstrate considerable resilience in the face of degradation. If some connections are removed or if noise is added to them, there is still a possibility that useful information may be retrieved from the system – again, a characteristic shared by real brains.

Connectionism has had an enormous impact in the last decade across the whole field of human learning, from perceptual-motor learning to language acquisition. In this chapter we will consider in detail how connectionist models operate and learn, and ask how successful they are at explaining the basic phenomena of human associative learning.

The delta rule

While there exist a number of connectionist learning rules, in this chapter I shall focus on one of the simplest, called the 'delta rule'. This rule has played a major role in several recent connectionist models of human learning. It was first described by Widrow and Hoff (1960), and is formally equivalent – given certain assumptions (Sutton and Barto, 1981) – to the theory Rescorla and Wagner (1972) developed to account for data from animal conditioning experiments.

Suppose we have a large set of potential cues or actions and an equally large number of possible outcomes or categories. We assume that each of the cues is represented by a unit in a homogeneous input layer of a large, highly interconnected network such as that shown in the left panel of Figure 4.1. Each output is also represented by a unit in a separate output layer, and each input unit is connected to each output unit with a modifiable connection (for the moment, we ignore possible hidden units).

Networks of this form are called *feedforward* networks or *pattern associators*. In reality, each cue and each outcome would be represented not by a single unit, but by a large collection of units corresponding to the elements (or microfeatures) of the stimulus, and the subject would have to learn structural associations amongst the elements of the cue, amongst the elements of the outcome, as well as the association between the cue and outcome. We will see below how this may occur, but I assume for the moment that all of the within–cue and within–outcome associations have already been learned and that these stimuli can therefore be represented by single units in the network.

On every trial, some set of cues is present and some outcome occurs. We calculate the activation a_o of each output unit:

$$a_o = \sum_i a_i w_{io} \, , \tag{4.1}$$

where a_i is the activation of input unit i and w_{io} is the current weight from cue unit i to output unit o. For a binary-valued cue, a_i will be 1 if that cue is present and 0 if it is not, whereas for a continuous-dimension cue, a_i will take on a value corresponding to the value of the cue on that dimension. Next, we calculate the 'error,' d_o, on the output unit between the obtained output, a_o, and the desired output, t_o:

$$d_o = t_o - a_o. \tag{4.2}$$

Values of t_o for binary and continuous dimensions are determined in a similar way to values of a_i, and represent the feedback provided to the learner. Finally, we change each of the weights in proportion to the error:

$$\Delta w_{io} = \alpha \, a_i \, d_o, \tag{4.3}$$

where Δw_{io} is the weight change and α is a learning rate parameter for the cue.

At the end of a series of training trials, there are a variety of ways of looking at the system's behavior. If we are modelling a situation in which subjects make cue–outcome association judgments, then the appropriate measure from the network is simply the activation a_o of the appropriate outcome unit when the input unit corresponding to that cue is turned on. On the other hand, if we are interested in classification performance, we can assume that the probability of classifying the stimulus as a member of category j will simply be the output a_j produced on the category j output node given cue c divided by the total output across all n output units:

$$P(j/c) = \frac{a_j}{\sum_n a_n} \tag{4.4}$$

Turning now to structural associations within a single stimulus, how might these be learned? An artificial grammar learning experiment provides a model for such within-stimulus learning. For a learning task like this, it is appropriate to use a somewhat different connectionist architecture known as an *autoassociator*, and the right-hand panel of Figure 4.1 shows such a network. A single set of units is activated by an input stimulus. Activation then spreads through the network on the weighted connections to produce an output pattern which is compared with the input pattern. Although the architecture is somewhat different, the system is governed by exactly the same equations as before. That is, each unit sums the activation it receives (Equation 4.1), compares this with the teacher, which in this case is the input activation, and derives an error (Equation 4.2). The error then determines weight changes in accordance with the delta rule of Equation 4.3.

Using one type of network for between-stimulus associations and another for within-stimulus associations may seem unjustified, but in fact the differences between feedforward and autoassociative networks are more apparent than real. An autoassociator can be considered as a feedforward network in which the input $(a_{i1}, a_{i2}, a_{i3},...)$ and output patterns $(a_{o1}, a_{o2}, a_{o3},...)$ are identical and where the connections between an input unit and the corresponding output unit $(i_1-o_1, i_2-o_2,$ etc.) are deleted. Of course, if such connections existed, then learning the mapping from an input pattern to an identical output pattern would be trivial: all that would be required would be to set each of these weights to 1.0. The removal of these connections means that the network is forced to predict a given element of the output pattern on the basis of different elements of the input pattern, such that element o_2 has to be predicted on the basis of elements $i_1, i_3, i_4,...$ and so on. The system is therefore forced to learn about the internal structure of the set of elements making up the cue.

Despite the formal similarity between feedforward and autoassociative networks, it is worth keeping them separate because the tasks to which they are best suited tend to differ. The only significant respect in which the behaviour of an autoassociator differs from that of a pattern associator is that it can be allowed to cycle for several iterations before it settles. The activation of a unit after one pass will lead to changes in the activation it transmits to other units, which will in turn lead to a change in its own activation, and so on until the system settles into a stable state. Therefore, in some applications it is possible to take the number of cycles needed to reach a stable state as a measure of response latency.

As described above, the procedure for training a connectionist network is to provide a set of input patterns together with their associated target patterns. As a result of incremental weight changes dictated by the delta rule, the network will come to produce the correct output pattern for each input pattern. It is important to note that the delta rule is guaranteed (given enough trials) to produce a set of weights that is optimal for the training stimuli (see Kohonen, 1977). That is to say, the rule will find a set of weights that minimises the squared error between actual and desired output patterns. Often, this squared error will be zero, meaning that the network produces exactly the desired output for each input pattern. We will discuss below some cases where this is impossible unless extra units are included between the input and output units, but even without such units, the delta rule will minimise the squared prediction error. The ability to prove that a learning rule will *converge* in this way towards an acceptable solution is of great benefit.

The connectionist theory outlined above has been applied in various forms to a very large number of associative learning tasks, ranging from the very simple to the very complex. I begin by showing that a network using the delta rule as its learning algorithm effectively computes the conditional degree of contingency between a cue and an outcome.

The computation of contingency

Let us suppose that subjects in an experiment are exposed to intermixed $AB \rightarrow O_1$ and $B \rightarrow$ no O trials in a contingent condition, while in a noncontingent condition they see $CD \rightarrow O_2$ and $D \rightarrow O_2$ trials. Cues B and D could either be explicit cues or just the constantly-present set of background cues. We know from the Shanks (1991a) experiment (Table 2.1) that the learned relationship between cue A and outcome O_1 will be much stronger than that between C and outcome O_2. According to the delta rule, the explanation of this phenomenon is as follows.

Imagine a network consisting of two input units corresponding to the two cues and one output unit for the outcome. Let us also simplify matters by assuming that the AB and B trials alternate and that the learning rate parameter α which appears in Equation 4.3 is 0.5 for all cues. On the first $AB \rightarrow O_1$ trial in the contingent condition, the outcome is entirely unexpected since the weights for all cues are zero. Hence a_o is zero while the teacher t_o has a value of 1.0. From Equation 4.2 we can calculate that d_o will be 1.0, and hence the weight change for each cue that is present – in this case both A and B – will be

$$\Delta w_{io} = \alpha \, a_i \, d_o$$
$$= 0.5 \times 1.0 \times 1.0$$
$$= 0.5.$$

Thus each cue ends the trial with a weight of 0.5. Trial 2 is a $B \rightarrow$ no O trial. Here, Equation 4.1 tells us that a_o will be 0.5 since the only cue present is B, and its weight is 0.5. The teacher t_o is zero since the outcome does not occur, meaning that d_o will be –0.5. Thus we obtain a weight change for each cue that is present – in this case just cue B – of:

$$\Delta w_{io} = \alpha \, a_i \, d_o$$
$$= 0.5 \times 1.0 \times (-0.5)$$
$$= -0.25$$

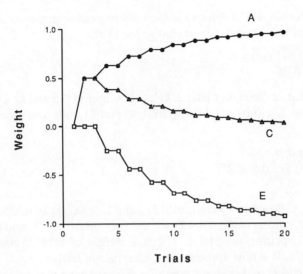

Figure 4.2. The ability of the delta rule to account for contingency effects. The figure shows the weights generated by the delta rule for various stimuli. Cue A has a positive contingency with outcome O_1, resulting from intermixed $AB{\rightarrow}O_1$, $B{\rightarrow}$no O trials, and the weight on the $A{-}O_1$ connection increases steadily across trials towards a value of 1.0. Cue C is noncontingently related to outcome O_2 as a result of $CD{\rightarrow}O_2$, $D{\rightarrow}O_2$ trials. The $C{-}O_2$ weight approaches zero after a number of trials. Cue E has a negative contingency for outcome O_3, resulting from $EF{\rightarrow}$no 0, $F{\rightarrow}O_3$ trials. The $E{-}O_3$ weight is reduced across trials towards an asymptote of -1.0.

which means that at the end of trial 2, A has a weight of 0.5 and B a weight of 0.25. On trial 3, an $AB{\rightarrow}O_1$ trial, the same logic dictates that A and B both acquire an increment to their weights of:

$$\Delta w_{io} = \alpha\, a_i\, d_o$$
$$= 0.5{\times}1.0{\times}0.25$$
$$= 0.125$$

since a_o is 0.75. Cue A now has a weight of 0.625 and B a weight of 0.375. After a further $B{\rightarrow}$no O pairing on trial 4, B's weight will have gone down to 0.1875, and after a further $AB{\rightarrow}O_1$ trial on trial 5, A will have a weight of 0.7188 and B a weight of 0.2813. It should be apparent that each successive pair of trials leaves A with a greater weight and B with a smaller weight, a process that will continue until A and B have weights of 1.0 and 0.0 respectively. Figure 4.2 shows how the weight of a contingent cue evolves across 20 trials under these conditions. The speed of learning is governed by the magnitude of α, the learning rate parameter.

Turning now to the noncontingent condition, a similar set of calculations reveals that in contrast to cue A, cue C will end up with a weight of zero, as required. The first $CD{\rightarrow}O_2$ trial is logically identical to the first $AB{\rightarrow}O_1$ trial and hence the weight change for each cue will be 0.5. Trial 2 is a

$D \rightarrow O_2$ trial. Now, a_0 is 0.5 since D is the only cue occurring, d_0 is also 0.5 ($=1.0 - 0.5$), and we obtain a weight change for D of:

$$\Delta w_{io} = 0.5 \times 1.0 \times 0.5$$
$$= 0.25$$

which means that at the end of trial 2, C has a weight of 0.5 and D a weight of 0.75. On trial 3, a further $CD \rightarrow O_2$ trial, C and D both lose associative strength:

$$\Delta w_{io} = \alpha \, a_i \, d_o$$
$$= 0.5 \times 1.0 \times (-0.25)$$
$$= -0.125$$

since a_0 is 1.25. C now has a weight of 0.375 and D a weight of 0.625. After a further $D \rightarrow O_2$ pairing on trial 4, D's weight goes up to 0.8125, and after a further $CD \rightarrow O_2$ pairing on trial 5, C has a weight of 0.28125 and D a weight of 0.71875. It is now apparent that each pair of trials leaves C with a smaller weight and D with a larger weight, a process that will continue until C and D have weights of 0.0 and 1.0 respectively. The net effect is the opposite of that in the contingent case, and the continued evolution of the weight of a noncontingent cue is shown in Figure 4.2. Note that after five trials (i.e., at the beginning of trial 6), the contingent cue A had a weight of 0.7188 while the noncontingent cue C has a weight of 0.28125, despite the fact that they have both been paired with their respective outcomes three times. Plainly, what happens on trials when a cue is absent may indirectly affect its associative strength.

Finally, Figure 4.2 shows the development of the weight for a cue in an $EF \rightarrow$ no O, $F \rightarrow O_3$ design in which the target cue E stands in a negative relationship to outcome O_3. For this cue, the weight starts at zero and then decreases in a negatively-accelerated manner towards a final value of -1.0. Plainly, the delta rule yields appropriate asymptotic weights under the three different contingencies we have considered.

We know from Chapter 2 that associative learning (within certain constraints) is based on the computation of contingency as defined by the ΔP equation and the probabilistic contrast model. For a situation in which there is a single cue (or action) occurring against a constant background, Chapman and Robbins (1990) have established the very important fact that the delta rule, at asymptote, yields weights that are identical to the values of ΔP. The proof of this is very elegant and well worth briefly examining. If we consider a set of AB and B trials, with A as the target cue, then there are four possible trial types ($AB \rightarrow O$, $AB \rightarrow$ no O, $B \rightarrow O$, and $B \rightarrow$ no O) bearing in mind that the outcome can either occur (O) or not occur (no O) on a given trial. Remember that we refer to the four cells of the 2×2 contingency table by the cell frequencies a ($AB \rightarrow O$, target cue and outcome present), b ($AB \rightarrow$ no O, cue present/outcome absent), c ($B \rightarrow O$, cue absent/outcome present), and d (B→no O, cue and outcome absent). We also specify that N is the total number of trials (= $a+b+c+d$). Hence the probability of a trial occurring with the target cue and the outcome is a/N.

For these various trial types the weight changes can be computed from Equations 4.1–4.3 as follows:

AB→O

$$\Delta w_B = \alpha_B [1.0-(w_A+w_B)] \tag{4.5}$$
$$\Delta w_A = \alpha_A [1.0-(w_A+w_B)] \tag{4.6}$$

AB→no O

$$\Delta w_B = \alpha_B [0.0-(w_A+w_B)] \tag{4.7}$$
$$\Delta w_A = \alpha_A [0.0-(w_A+w_B)] \tag{4.8}$$

B→O

$$\Delta w_B = \alpha_B (1.0-w_B) \tag{4.9}$$

B→no O

$$\Delta w_B = \alpha_B (0.0-w_B) \tag{4.10}$$

where w_A refers to the weight connecting cue A with outcome O, and α_A and α_B are learning rate parameters for A and B, respectively.

Across a series of intermixed trial types, the mean change in cue A's weight is the sum of the weight changes given in Equations 4.6 and 4.8, weighted by the probabilities (a/N and b/N) of these two trial types:

$$Mean\ \Delta w_A = \frac{a}{N}\ \alpha_A\ [1.0 - (w_A + w_B)] + \frac{b}{N}\ \alpha_A\ [0.0 - (w_A + w_B)] \tag{4.11}$$

which can be rewritten as:

$$Mean\ \Delta w_A . \frac{N}{\alpha_A} = a - (a + b).w_A - (a + b).w_B \tag{4.12}$$

Similarly, the mean change in cue B's weight is the sum of the weight changes in Equations 4.5, 4.7, 4.9, and 4.10 multiplied by the probabilities a/N, b/N, c/N, and d/N of the different trial types:

$$Mean\ \Delta w_B = \frac{a}{N}\ \alpha_B[1.0 - (w_A + w_B)] + \frac{b}{N}\ \alpha_B\ [0.0 - (w_A + w_B)]$$
$$+ \frac{c}{N}\alpha_B\ (1.0 - w_B) + \frac{d}{N}\alpha_B(0.0 - w_B) \tag{4.13}$$

which can be rewritten as:

$$Mean\ \Delta w_B . \frac{N}{\alpha_B} = a + c - (a + b).w_A - (a + b + c + d).w_B \tag{4.14}$$

Learning will proceed until the mean weight changes given by Equations 4.12 and 4.14 across a sequence of trials have reached zero. That is to say, even though at asymptote the weights may fluctuate from trial to trial if the occurrence of the outcome is determined probabilistically, they will tend to oscillate around a fixed value such that the *mean* change across a number of trials is zero. Setting Equations 4.12 and 4.14 at zero and substituting for w_B, we find that:

$$0 = a - (a+b)w_A - \frac{(a+b)[a+c-(a+b)w_A]}{a+b+c+d}$$

which when tidied up yields:

$$w_A = a/(a+b) - c/(c+d) = \Delta P.$$

Chapman and Robbins' proof establishes, then, that the weight assigned by the delta rule to a cue will be equal to the degree of statistical contingency between the cue and the outcome. The proof also shows that the values of the learning rate parameters α_A and α_B do not alter the asymptotic weights, merely the speed with which asymptote is reached. However, it is important to emphasise that the equivalence of ΔP and the delta rule only holds at asymptote. Prior to asymptote, a cue's weight may be very different from ΔP. This is important, because as we shall see, it means that we can explain many of the phenomena (such as the shapes of learning curves) that appeared to be problematic for contingency theory.

As would be expected given this proof, connectionist models have been able to provide excellent fits to data from associative learning experiments varying the degree of contingency between events. For example, Wasserman *et al.* (1993) obtained good fits to the data presented in Figure 2.3. Remember that Wasserman et al. asked subjects to estimate the degree of contingency between pressing a key and a light flashing. Across 25 different problems, P(O/A) and P(O/–A) independently took the values 0.0, 0.25, 0.50, 0.75, and 1.00. Wasserman *et al.* fit the delta rule model to their observed data, and the outcome is shown in Table 4.1. As the table shows, a very good fit is obtained. Note that the predicted values are slightly pre-asymptotic, which explains why they are not equal to the expected values of ΔP.

The delta rule not only yields weights that converge to appropriate asymptotic values, but it also produces learning functions similar to those obtained in associative learning tasks. The left-hand panel of Figure 4.3 reproduces the data from the experiment by Lopez and myself that was discussed in Chapter 2 (Figure 2.4). Recall that we found that under a positive contingency ($\Delta P=0.5$), judgments started close to zero and increased in a negatively-accelerated manner until they reached an asymptote that corresponded approximately to the actual contingency. Under a negative contingency ($\Delta P=-0.5$), judgments again started close to zero and decreased towards asymptote. The delta rule predicts exactly such functions. Since the w_{io} start at zero, the initial errors d_o will be large and hence increments to the

Table 4.1. *Results of Wasserman et al.'s (1993) experiment and simulation*

	P(O/–A)				
	0.00	0.25	0.50	0.75	1.00
P(O/A)					
1.00	0.85	0.52	0.37	0.13	–0.03
	0.80	0.51	0.30	0.13	0.00
0.75	0.65	0.43	0.16	0.06	–0.12
	0.67	0.38	0.17	0.00	–0.13
0.50	0.37	0.19	0.01	–0.19	–0.34
	0.50	0.21	0.00	–0.17	–0.30
0.25	0.12	–0.10	–0.14	–0.37	–0.58
	0.29	0.00	–0.21	–0.38	–0.51
0.00	–0.08	–0.45	–0.51	–0.66	–0.75
	0.00	–0.29	–0.50	–0.67	–0.80

First row of each pair: mean judgments of contingency (divided by 100.0) as a function of P(O/A) and P(O/–A). Second row: associative strengths predicted from the delta rule.

weights will in turn be large, but successive increments will get smaller and smaller as d_o decreases. The net effect will be negatively accelerated acquisition curves. The right-hand panel of Figure 4.3 shows a set of learning curves generated for the same contingencies used in our experiment. The model predicts the correct terminal pattern of judgments and reproduces the acquisition profiles under the different contingencies, namely negatively accelerated curves for the high positive and high negative contingencies.

Interestingly, the figure shows that the delta rule also reproduces quite well the patterns of judgments seen in noncontingent conditions. For both the high-frequency (0.75/0.75) and low-frequency (0.25/0.25) noncontingent situations, the weights converge to zero, as in the judgments shown on the left. Moreover, prior to asymptote, the rule correctly yields a substantially more positive weight in the 0.75/0.75 condition than in the 0.25/0.25 condition. Why does it do this? In fact, we have already seen the answer. Our derivations for a noncontingent cue C in a $CD \rightarrow O_2$, $D \rightarrow O_2$ design produced a positive weight early on in training (see Figure 4.3). The reason is that on the first few trials, before D has begun to acquire a significant weight, C is paired with the outcome and thus acquires a positive weight. It

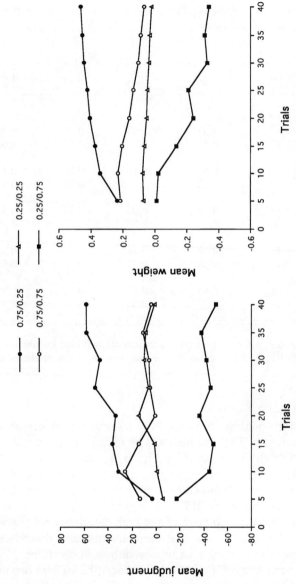

Figure 4.3. Acquisition data and simulation. The left panel is a reproduction of Figure 2.4, and shows mean judgments of contingency across 40 trials under four different contingencies: positive (0.75/0.25), zero (0.75/0.75 and 0.25/0.25), and negative (0.25/0.75), where the first figure refers to P(O/C) and the second to P(O/–C). The right-hand panel shows the results of a simulation using the delta rule. Each simulation curve is averaged across four separate runs with different trial orders. The learning rate parameter α was set to 0.2.

is only later, as the weight for cue D begins to increase, that it starts to capture associative strength back from C. When the overall frequency of the outcome is high (as in the 0.75/0.75 condition), the outcome occurs on most CD trials, and so C's weight increases quite fast. In contrast, in the 0.25/0.25 condition there are many CD trials where the outcome is absent, so C loses any associative strength it might otherwise gain.

In addition to the differences that may emerge between the delta rule and contingency models in their pre-asymptotic predictions, they may also make very different predictions when trial types are presented in blocks. Bear in mind that the proof of the equivalence of the two theories only holds for situations in which trials are intermixed. Let us consider some data that Chapman (1991) reported on the effects of trial order. It may be recalled from Chapter 2 (Table 2.8) that, using a medical diagnosis task, Chapman presented subjects with A→O followed by AB→no O trials, and observed strongly negative ratings of cue B. The delta rule accounts for this effect on the basis that during the second stage, B is paired with a positive cue but the expected outcome fails to occur: B therefore acquires a negative connection. In her experiment, Chapman also gave subjects CD→no O followed by C→O trials, and since these trials are logically identical to the A and AB trials, except for the order of presentation, it may seem surprising that cue D received significantly less negative ratings than cue B.

In Chapter 2 it was pointed out that this result presents a problem for the view that associative judgments are based on a computation of statistical contingency. However, the result can be explained by a connectionist model. Applying the delta rule to the trials types, we can see that the CD→no O, C→O procedure should not have been able to endow cue D with negative strength, since it was accompanied in the first stage by a cue with zero strength and was never paired with the outcome. But note that this prediction relies on the fact that the trial types were not intermixed. According to a connectionist model, different solutions are possible when exactly the same trial types are witnessed: in the A→O, AB→no O case, the stable solution is for A to have a weight of 1.0 and B to have a negative weight (–1.0), whereas in the CD→no O, C→O case the stable solution assigns D a weight of zero and C a weight of 1.0. It is the fact that the trials were not intermixed that allows differential predictions to emerge.

Associative models using the delta rule learning algorithm can explain a further interesting phenomenon, namely the fact that associative learning is influenced by the magnitude or value of the outcome. Chatlosh *et al.* (1985) observed higher judgments and response rates under a positive contingency with a higher than with a lower valued outcome, while for negative contingencies, more valued outcomes seemed to support more negative judgments. Changes in the nature of the outcome have no effect, of course, on the statistical relationship between the cue or action and outcome. However, the delta rule model has an elegantly simple explanation of these results, since it proposes that the teacher t_o is proportional to the magnitude or value of the outcome: more valued outcomes have larger t_os. Since t_o determines the asymptote of associative strength (learning ceases when the associative strengths sum to t_o in Equation 4.2), an outcome with a large t_o will support greater judgments and response rates under a positive contingency than will an outcome with a smaller t_o, just as was found by Chatlosh *et al.* (1985). For negative contingencies, more valued outcomes will support more negative judgments.

One limitation of Chapman and Robbins's proof is that it only covers situations involving a single target cue and the background. As we saw in Chapter 2, though, the appropriate way to compute contingency when there are multiple cues is specified by the probabilistic contrast model. For a given target cue, we need to calculate the probability of the outcome given the cue, minus the probability of the outcome in the absence of the cue, but holding everything else constant. Can we determine whether the delta rule will compute associative strengths in accordance with the contrast model?

In a very important theorem, Cheng and Holyoak (1995) have proved that the delta rule, within certain constraints, will behave in this fashion. Their proof is complex and will not be presented here, but it establishes that at asymptote the associative strength of a cue will equal the value of ΔP as determined by the appropriate probabilistic contrast. For example, in a design with trial types A, AB, and ABC, with some of these trials being paired on some occasions with the outcome, the asymptotic weight for cue C will be equal to P(O/ABC) – P(O/AB), and the weight for cue B will be P(O/AB) – P(O/A). In each case, the weight corresponds to ΔP defined over the appropriate contrast: for cue C, for example, the contrast is evaluated across the ABC and AB trials.

What are the limitations of this proof, other than that training be continued to asymptote? Cheng and Holyoak were only able to establish the proof for situations in which the trial types are nested in a particular way. Specifically, the proof only applies when the trials are such that every cue combination (such as ABC) is a proper superset of each smaller combination (such as A and AB). According to this restriction, a design involving trial types such as A, AB, and BC would not be acceptable because BC is not a superset of the smaller set A. In cases such as this, the probabilistic contrast model is inapplicable anyway since the relevant contrast (say for cue C) cannot be computed, so the question of whether its predictions are identical to those of the delta rule does not arise. At any rate, the proof has sufficiently wide applicability to encourage the view that the two theories will yield equivalent asymptotic predictions in all but a few situations.

Given Cheng and Holyoak's theorem, it is perhaps not surprising that the delta rule model can explain many of the selective learning effects that were described in Chapter 2 and which could be understood in terms of conditional contrasts. Let us first consider the phenomenon of blocking.

Remember that Chapman and Robbins (1990) gave subjects A→outcome and B→no outcome trials prior to AC→outcome and BD→outcome trials (see Table 2.5). According to the delta rule, in the first stage cue A comes to acquire a positive weight while the B-outcome weight has zero strength. In the second stage, on an AC trial the weight of cue A will make the term a_o large in Equation 4.1, while on a BD trial a_o will in contrast be zero. Hence the mismatch d_o will be smaller for C than for D, and the weight change in Equation 4.3 will in turn be smaller for C than D. As the second stage of the experiment proceeds, Δw_{io} will always be smaller for cue C than for cue D and hence the weight will never attain such a high value.

The effect of signalling noncontingent outcomes (see Figure 2.5 in Chapter 2) is explained in a similar way. Remember that associative judgments are increased in an action–outcome situation if an additional cue signals those outcomes that are response-independent. The effect occurs, according to the theory, because the presence of the signal as an additional cue prevents the context or background from acquiring positive associative strength, and hence from competing with the action on action–outcome trials, as it does in the noncontingent control condition. Essentially, the signal blocks the background which in turn is less able to block the action. Let us look at the calculations in more detail.

If we have AB→O and B→no O trials then the target cue (A) is contingently related to the outcome, and when we have AB→O and B→O trials the outcome is not contingent on cue A. A signal condition is identical to a noncontingent one except that the noncontingent outcomes which occur in the absence of the target cue are accompanied by the signal, which means we have AB→O and BC→O trials types with cue C being the signal. We have already seen how weights evolve in the contingent and noncontingent conditions, so let us now run through the equations for the signal condition, with the same parameter values as before, and again assuming alternating trial types. On the first AB→O trial each cue will receive a weight increment of 0.5 as before. On the second trial, B and C are paired with the outcome. Since B has a weight of 0.5, d_o will be 0.5 and the change in the weights for B and C will be

$$\Delta w_{io} = \alpha \, a_i \, d_o$$
$$= 0.5 \times 1.0 \times 0.5$$
$$= 0.25$$

which means that B will have a total weight of 0.75 and C a weight of 0.25. On the next AB→O trial, d_o is –0.25 [= 1.0 – (0.5+0.75)] and so A and B will each lose some associative strength:

$$\Delta w_{io} = \alpha \, a_i \, d_o$$
$$= 0.5 \times 1.0 \times (-0.25)$$
$$= -0.125,$$

so A now has a weight of 0.375. After a further BC→O trial, B has a weight of 0.6875 and A a weight of 0.375. Another pair of trials ends with A having a weight of 0.3437. At a comparable point in the noncontingent condition, A had a weight of 0.2813, so it can be seen that A's weight is greater in the signal than in the noncontingent condition. In fact, with the parameters used here, A's asymptotic weight is 0.33, whereas the corresponding weight for a noncontingent cue would of course be zero. What is essentially happening is that cue C forces cue B to acquire a weight of less than 1.0. This in turn means that B is unable to force A's weight all the way down to zero. In sum, the calculations show that the effect of a cue signalling noncontingent occurrences of the outcome is to augment the weight of the target cue, just as we require.

A final basic learning effect that can be readily explained by a connectionist model comes from another of the experiments (Shanks, 1991a) considered in Chapter 2. In that experiment, I used the medical diagnosis procedure and asked subjects to rate the relationship between cue A and outcome O_1 and that between cue D and outcome O_2 after the training trials shown in Table 4.2. The procedure again required subjects to make diagnoses on each trial, with corrective feedback. For outcome O_1, the problem is easily mastered since this disease only occurs on AB trials; AC trials were accompanied by the absence of any disease. For outcome O_2 the task is ambiguous since both DE and DF were paired with this disease on 50% of trials in the learning task.

The table also shows the weights obtained in a feedforward network in which each symptom is represented by a separate input unit, and where there were two output units corresponding to the two diseases. On trials where no disease occurred, the teaching signal consisted of activations of zero for each output unit. The network was trained for the same number of trials as the subjects. As can be seen, the model reproduces the finding of higher judgments for cue D in the uncorrelated than for cue A in the correlated condition. The reason for the difference in the ratings is that in the correlated condition, cue B is able to block the acquisition of associative strength by cue A.

To summarise where the present section has led us: the delta rule can be shown to compute ΔP at asymptote, both when the background is constant and when it is variable. In the latter case, the rule yields weights in correspondence with the probabilistic contrast model. Accordingly, the main findings discussed in Chapter 2 can all be reproduced by an associative system in which weights are updated by the delta rule, and to this extent, associationist learning mechanisms are able to explain why learning is normative.

Table 4.2. *Design and results of the experiment by Shanks (1991a, Experiment 3)*

Condition	Trial types	Test symptom	Mean rating	Weight
Correlated	AB→O_1 AC→no O	A	32.3	33.5
Uncorrelated	DE→O_2 DE→no O DF→O_2 DF→no O	D	49.0	59.3

A–F are the cues (symptoms) and O_1, O_2 are the outcomes (diseases); no O indicates no outcome.

Representation in connectionist networks

In the last chapter we saw that an instance theory like the context model can account for much of the data obtained in concept learning tasks. Recall that there are three principal phenomena for a categorisation model to account for, (i) rapid and accurate responding to an unseen prototype, (ii) evidence that categorisation is mediated at least to some degree by memorised instances, and (iii) the base-rate results of Gluck and Bower (1988) and Shanks (1990). How do connectionist models fare with these phenomena? In particular, to what extent can it be said that connectionist systems memorise training stimuli? I shall deal first with prototype and instance effects before turning to the base-rate data.

Beginning with prototype effects, it has been known since the seminal articles by Knapp and Anderson (1984) and McClelland and Rumelhart (1985) that connectionist networks are able to reproduce these effects. This comes about simply because the prototypical pattern activates all of the input units which are most strongly associated with the target category, and few of the units that are strongly associated with other categories. The matrix of weights the network acquires during training represents an abstraction from the training exemplars corresponding (in a very loose way) to a prototype. Moreover, it is this abstraction process that allows connectionist networks to manifest the same sort of behaviour as Watkins and Kerkar's (1985) subjects. Recall (see Figure 3.14) that exposure to two instances (say of the word 'umbrella') made the word itself much easier for subjects to remember but made the colours of each presentation harder to remember. Because the word itself is constant across presentations, a connectionist system such as an autoassociator will tend to assign considerable weight to the features of the word. In contrast, because the colour varies, that attribute will be to some extent discounted. The net effect will be a degree of loss of idiosyncratic information from the weight matrix.

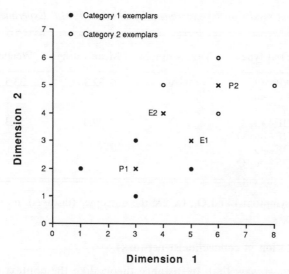

Figure 4.4. Locations of hypothetical stimuli to illustrate the ability of the delta rule to reproduce instance effects. The stimuli are assumed to vary on two dimensions, with P1 and P2 being the prototypes of categories 1 and 2, respectively. A feedforward network with two output units (one for each category) was trained to classify the four training exemplars from each category and was then tested with the prototypes and the critical test stimuli E1 and E2. Despite the fact that E1 and E2 are equidistant from the prototypes, E1 was assigned to category 1 because it was more similar to a training stimulus from that category, and E2 was assigned to category 2 because it was more similar to a training stimulus from that category. (After Shanks, 1991b.)

What about instance effects? On the face of it, such effects should be difficult to reproduce since each weight in the network is the result of experience of a very large number of different exemplars. One of the clearest examples came from the experiment by Homa *et al.* (1981). To try to reproduce such effects, at least in essence, I (Shanks, 1991b) presented a network with four hypothetical training stimuli from each of two categories, and then tested it with new stimuli equidistant from the prototypes but differing in similarity to specific training items. The arrangement of the stimuli is shown in Figure 4.4.

Each stimulus was represented by a pattern of activation across 24 input units using a coding scheme that ensured that stimuli close together were represented by similar patterns of activity. Each of the eight training stimuli was presented six times to the network with appropriate feedback, and then responding was measured to the prototypes and to the critical test stimuli E1 and E2. Note that E1 is equidistant from the prototypes but slightly more similar to a category 1 exemplar than to the nearest category 2 exemplar, whilst E2 is also equidistant from the prototypes but is slightly more similar to a category 2 than a category 1 exemplar. Confirming that this sort of model can produce instance effects, the network responded differently to the two test stimuli, classifying E1 as a member of category 1 and E2 as a member of category 2, despite the fact that they were equidistant from the prototypes. The reason for this is that although the weights in the network are abstractions in the sense that they combine information across multiple presentations of different stimuli, they nevertheless do implicitly retain some information about specific training instances. Each instance possesses some idiosyncratic features, and although such features will only be seen infrequently, the network will nevertheless be able to learn that such features are weakly predictive of the category. Two stimuli equidistant from the prototype may then differ in terms of possession of such weakly-predictive idiosyncratic features.

Prototype and instance effects are also reproduced by autoassociative networks. The usefulness of such systems for understanding associative learning can be appreciated by considering some simulations reported in a classic article by McClelland and Rumelhart (1985) which paved the way for a wealth of subsequent investigations of the correspondences between human category learning and learning in connectionist models. McClelland and Rumelhart showed that an autoassociative network can reproduce the instance results obtained by Whittlesea (1987). Remember that Whittlesea's subjects saw letter strings (see Table 3.4) such as FUKIP and attempted to learn something about the internal structure of the strings. What are the rules that govern the formation of the letters? In one experiment subjects were trained on the IIa items and tested on the IIa, IIc, and III items. Although the IIa and IIc items are equidistant from the prototype, subjects showed more facilitation to the IIa than the IIc items as shown in Table 4.3. Also, the IIc items are closer to the prototype that the III items, yet the latter showed more facilitation, which Whittlesea attributed to the fact that the III items were more similar to the studied IIa items than were the IIc items.

McClelland and Rumelhart (1985) simulated these results by presenting a 20-unit autoassociative network with exactly the same events as subjects. Each unit coded a letter occurring in a given position. In the learning stage, a stimulus was presented on each trial and weights adjusted according to the delta rule, and then in the test stage the relevant test items were presented to the network (with learning switched off). Table 4.3 shows that both of the key results were reproduced by McClelland and Rumelhart's model, where the figures represent the increase in the dot product between the input and output vectors. In showing a greater degree of facilitation to the IIa than to the IIc items the model is clearly demonstrating its ability to maintain information about the specific instances that it was trained with. The benefit of the III items over the IIc items shows that the network can respond to new test items as a function of their similarity to studied items.

Despite this ability to encode instances, in further simulations

Table 4.3. *McClelland and Rumelhart's (1985) simulation of Whittlesea's (1987) results*

Training stimuli	Test stimuli	Observed transfer	Predicted transfer
IIa	IIa	1.07	1.45
	IIb	0.80	0.70
	IIc	0.51	0.60
IIa	IIa	1.22	1.45
	IIc	0.65	0.60
	III	0.86	0.75
Ia	P	—	1.40
	Ia	—	1.20
	IIa	—	0.60

Observed transfer = mean increase in the number of letters correctly identified, relative to the baseline stage. Predicted transfer = increase in dot product between input and output vectors.

McClelland and Rumelhart established that their model can also show the superior responding to an unseen prototype that, as we saw in Chapter 2 in the experiment of Homa *et al.* (1981), is a feature of human classification. McClelland and Rumelhart trained their network on the Ia items that are similar to the prototype, and then tested it on the prototype, the Ia items, and the IIa items. The results, given in Table 4.3, show a clear benefit for the unseen prototype over the studied Ia items. Although Whittlesea did not conduct the equivalent of this experiment with his subjects, the simulation results correspond to those obtained in other circumstances (such as in Homa *et al.*'s experiment) where the prototype is more accurately or rapidly classified than the training items. In sum, McClelland and Rumelhart (1985) demonstrated an extremely impressive correspondence between data obtained in associative learning tasks and the behaviour of their connectionist model.

Turning now to base-rate effects, recall that the crucial data obtained by Gluck and Bower (1988) and others came from experiments in which there were AB→1, AB→2, and A→1 training trials. Despite being equally paired with categories 1 and 2, subjects tended to classify cue B in category 2. It turns out that this effect is easily reproduced by a connectionist model. Gluck and Bower (1988), Nosofsky *et al.* (1992), and Shanks (1990) have all shown that this bias can be reproduced in a connectionist network. Thus while subjects in the Shanks (1990) study chose disease 2 with probability p=0.63, a connectionist network trained on an identical set of learning trials generated weights of 0.33 and 0.05 on the A→2 and A→1 connections, respectively, which would translate into a strong bias (p=0.87) for choosing category 2. The reason is basically the same as the explanation of simple contingency effects: cue A becomes strongly associated with category 1, and this means that on the AB→1 and AB→2 trials cue B must acquire a greater weight for category 2 in order to offset cue A's bias towards category 1.

Artificial grammar learning

Given the success of McClelland and Rumelhart's model in acquiring knowledge of the internal structure of Whittlesea's letter strings, it is perhaps not surprising that such a model can also be successfully applied to the learning of artificial grammars. Dienes (1992) has recently shown that artificial grammar learning can be well understood in terms of the operations of a simple autoassociator. He tested various slightly different learning rules and coding schemes. The network's task was to reproduce across the units the pattern that was actually presented.

Dienes used the grammar that was briefly discussed in Chapter 1 (Fig. 1.1) and which generated strings like MVT and VXVRXR. Since there were five letters in the grammar, and a maximum of six letters in each string, 30 units are sufficient to code each possible letter occurring at each possible position in the string. During the learning phase, a string was presented as a pattern of input activations (1s and 0s) across the units. At the end of each trial, weights were updated according to the delta rule. This meant that by the end of the training phase the network had learned various correlations between letters in different positions. For instance, a V in position 1 led to activation of the unit representing letter X in position 2, encoding the fact that in the training items any string that began with a V continued with an X. In contrast, an M in position 1 activated to some degree the units representing T and V in position 2, since both of these were legal continuations for a string commencing with M.

To test the network, at the end of the training phase Dienes presented it with new strings that were either grammatical or nongrammatical. As a measure of whether the network regarded a string as grammatical, Dienes used the cosine of the angle between the input and output vectors as a measure of how well the network was able to recreate the input pattern. The results showed that this autoassociator network could provide an excellent account of artificial grammar learning, not only being able to discriminate between new grammatical and nongrammatical strings, but also performing at the same overall level of accuracy as subjects, and producing approximately the same rank ordering of difficulty of the test strings. The network behaves in this way because the weights represent weak correlations that exist between letters at different positions in the grammar. When a grammatical string is presented, there is a good match between the sequence of letters in the string and the internally-generated expectation the network has concerning the sequence. For a nongrammatical string, the internally-generated prediction mismatches the structure of the string.

In the last chapter, we saw that much of the data obtained from artificial grammar learning experiments can be interpreted in terms of memorisation of studied grammatical strings: if a new test string is sufficiently similar to the memorised items, it is called 'grammatical'. Some evidence for this came from Vokey and Brooks's (1992) observation that when grammatical status and similarity to specific study exemplars are separated, both factors appear to contribute to performance. Vokey and Brooks constructed sets of test strings (see Table 3.6) such that some were similar to study items (i.e., differing by one letter from the most similar study string) while others were dissimilar (differing by more than one letter). Orthogonal to this, half the test items were grammatical and half were nongrammatical. Vokey and Brooks found that similarity to studied exemplars played a significant role, in that more close than far test items were called grammatical. This of course supports the view that instance memorisation plays a role in the learning of such artificial grammars, but Vokey and Brooks realised that it could not be the whole story: over and above similarity, grammatical status continued to be a contributing factor. More grammatical than nongrammatical test strings were called grammatical, even when equated for similarity to study items.

It appears that connectionist models may be able to explain both of these effects within a single mechanism. A connectionist model such as Dienes' does not literally, of course, memorise the strings it is presented, but instead extracts information about the regularities that exist between the elements of the study strings. Although such a system will, naturally, be sensitive to the degree of similarity between studied and test strings, it may also be sensitive to grammaticality: a grammatical test string will, by definition, be made up of relations between elements that existed in the set of study strings.

All of this suggests that a unitary explanation of Vokey and Brooks' data may be possible, in contrast to the dual explanation they suggested. Accordingly, Perruchet (1994) has shown that if the strings used by Vokey and Brooks are broken into their letter pairs and triplets, then the two factors of similarity and grammaticality can be subsumed by just one, namely number of studied fragments. 'Close' test items similar to training strings contain more studied letter pairs and triplets than 'far' ones which are dissimilar to training strings, and independently, grammatical test items contain more studied pairs and triplets than nongrammatical ones. Thus the data can be understood in terms of exactly the sort of information that connectionist models extract, namely contingencies between the elements of the studied items. Furthermore, such a model does seem to conform quite well to the sorts of knowledge subjects report. The evidence suggests that subjects are quite good at recalling the legal letter pairs that they observed in the study phase.

Internal representations

In the models we have been considering thus far, knowledge is represented in weighted connections. In autoassociators, these connections are between elements of the stimulus itself, while in pattern associators they are between elements of the cue and elements of the outcome. While the learning mechanism we have examined, the delta rule, has been successfully applied to many tasks, there is evidence to suggest that this 'elemental' representational assumption is inadequate. In learning to associate one pattern with another, for instance, it appears that in addition to learning direct associations between the outcome and the separate elements that make up the stimulus, intermediate representations of the stimulus can also be involved in associations with the outcome.

The inadequacy of the notion that the elements of the cue are directly associated with the outcome comes from the fact that humans can learn nonlinearly separable classifications. In single-layer networks, in which just one layer of modifiable connections exists between the input and output units, it is easy to see that the predicted outcome a_o must be a linear sum of the inputs. Consider a network consisting of two input units (denoted x and y) connected to one output unit, where the inputs and correct output t_o can take on values between 0.0 and 1.0, and where the network is trained to classify input patterns into one of two categories. Regardless of the weights, Equation 4.1 tells us that the output a_o must always be a simple linear sum of the activations (a_x and a_y) of input units x and y:

$$a_o = a_x w_{xo} + a_y w_{yo}.$$

It follows that the only types of classification such a system can learn are *linearly separable* ones in which the members of the two categories can be distinguished by a simple linear boundary. Specifically, for the delta rule model to learn a classification, it must be possible to construct a straight line in the x, y input space that exactly divides the stimuli into the correct categories. If such a line can be drawn, then there exist weights that will allow the model to produce greater outputs for members of one category than for members of the other category. The classification is solved by making one category response whenever the output is greater than a threshold and the other response whenever it is less.

However, people have no difficulty learning nonlinearly-separable classifications which the delta rule model we have been considering would be unable to master. Indeed, we have already seen at least two examples of this. One was cited in Chapter 1: humans and animals can readily learn discriminations in which two red stimuli are shown on some trials and reward depends on choosing the right-hand one, while on other trials, a pair of green stimuli is presented and reward is given for choosing the left-hand stimulus. Such a discrimination cannot be solved by networks of the sort considered thus far,

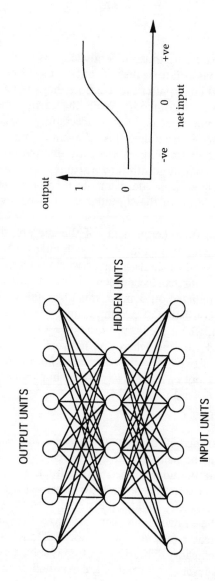

Figure 4.5. Left-hand panel: a backpropagation network. A homogeneous layer of input units is connected in parallel to a layer of hidden units which are in turn connected to a layer of output units. The connections between the units have modifiable weights. The backpropagation algorithm adjusts the input-hidden and hidden-output weights on a trial-by-trial basis until the correct pattern of activation is obtained on the output units. Right-hand panel: the logistic activation function. The graph shows the sigmoidal relationship between the net input to a hidden or output unit in a backpropagation network and the unit's output.

because each element (red, green, left, right) should be equally associated with reward. The second example comes from Nosofsky's (1987) categorisation data. In one of Nosofsky's classification tasks (the diagonal problem shown in Figure 3.8) the two categories cannot be separated by a linear discrimination, yet subjects had no obvious difficulty learning to solve that problem.

It is for this reason that many feedforward connectionist models incorporate a layer of 'hidden' units that intervene between the input and output units, as shown in Figure 4.5. Such a network can operate exactly like a single-layer pattern associator if we continue to use the delta rule as the learning algorithm. One particular type of hidden-unit network has been extremely widely investigated and has been shown to have some very powerful properties. In such a 'backpropagation-of-error' network, the delta rule applies exactly as before except that it is refined in order to determine how much the input-hidden weights and the hidden-output weights should be changed on a given trial. The precise calculations are as follows (for further details, see McClelland and Rumelhart, 1988). Each hidden unit h computes its activation, a_h, which is a logistic function of its input:

$$a_h = \frac{1}{1+e^{-\sum_i a_i w_{ih}}},$$

where w_{ih} is the weight connecting input unit i to the hidden unit and a_i is the activation of input unit i (1.0 if the relevant feature is present and 0.0 if it is absent). The reason for using this logistic function is that it is mathematically very convenient from the point of view of partitioning the weight changes between the input-hidden and the hidden-output connections.

Each output unit o computes its activation a_o in exactly the same way by summing its inputs times the weights from the hidden units and putting the total input through a logistic transform. Finally, the weights, which start out not at zero but with small random values, are changed according to Equation 4.3. For the weights connecting the hidden and output units, the relevant error term d_o is:

$$d_o = (t_o - a_o) \cdot a_o \cdot (1.0 - a_o), \tag{4.15}$$

where t_o is again the teaching signal on the output unit. For the input-hidden connection the relevant error is

$$d_h = a_h \cdot (1 - a_h) \cdot \sum_o w_{ho} d_o \ .$$

The development of multilayer networks using this generalised version of the delta rule has provided a major contribution to recent connectionist modelling since phenomena such as the learning of nonlinear classifications that are impossible for single-layer networks can be easily dealt with by multilayer networks. Although Minsky and Papert (1969) were quite correct in highlighting the many weaknesses of single-layer networks, it is regrettable that their critique led so many researchers to lose interest in associationist learning devices. The development of training procedures for multilayer networks may otherwise have occurred somewhat earlier.

Be that as it may, even more impressive than their ability to learn nonlinear classifications is the fact, proved by Hornik, Stinchcombe and White (1989), that backpropagation networks can learn essentially any mapping between a set of input and output patterns that one cares to construct. Thus, for any set of mappings from arbitrary input patterns to arbitrary output patterns ($I_1 \rightarrow O_1$, $I_2 \rightarrow O_2$, $I_3 \rightarrow O_3$,...), a backpropagation network with sufficient hidden units will construct a set of weights to learn the mapping to any desired degree of approximation. Hence there is no question about the power of this sort of connectionist network for learning associative relationships. But the question remains, does it learn in the same way as humans?

There is undoubtedly evidence of persuasive correspondences between human behaviour and the predictions of backpropagation networks. Some of the best evidence concerns child language acquisition, where it is possible to provide a network with approximately the same sort of input that children receive and see whether characteristics of the network's learning match those seen in children. One much-debated example concerns the learning of the past-tense in English. While most verbs are regular in adding -ed to produce the past tense (e.g., walk–walked), a number of very common verbs are irregular (e.g., go–went, send–sent, have–had, etc.). Children, of course, are able eventually to learn the correct past tenses, but they also produce some interesting errors in that they occasionally 'over-regularise' irregular verbs: they say 'goed,' 'sended', and so on. It turns out that backpropagation networks are also able to produce such errors (Plunkett and Marchman, 1991). Because they encounter many more regular than irregular verbs, early on in training the network may inappropriately generalise the contingency between verb stems and the -ed past tense and apply it to irregular verbs.

From the more general perspective of human associative learning, however, the basic backpropagation system is inadequate for at least two reasons. The first is that, contrary to the available evidence, such a system will learn a linearly-separable classification of a set of stimuli faster than a nonlinear partition of the same stimulus set. The reason for this is that hidden units in a backpropagation network can each be thought of as attempting to construct a line or plane in the input space to classify the stimuli appropriately. If there is indeed a linear separation of the stimuli, then it only takes one hidden unit to orient itself appropriately and the classification is solved. In contrast, a nonlinear classification requires more than one appropriately-aligned hidden unit, and the alignment of two or more units will never take less time than the alignment of one.

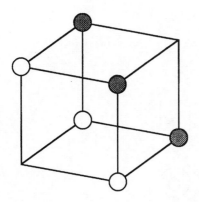 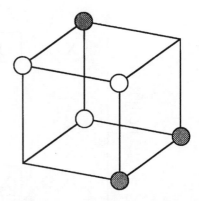

Figure 4.6. Stimuli used by Medin and Schwanenflugel (1981) to study learning of linearly-separable (left) and nonlinearly-separable (right) classifications. Stimuli varied on three binary-valued dimensions and the same set of stimuli was used in both cases. Subjects were taught to assign stimuli represented by open circles to one category and those represented by filled circles to the other. Note that a plane can be constructed to divide perfectly the two sets of stimuli for the linearly-separable classification, but this is not possible for the nonlinearly-separable one.

Not only can humans actually learn nonlinear classifications, but there is also evidence that under controlled conditions, they are learned at least as fast as linear ones. The best-known example is from Medin and Schwanenflugel (1981). In one of their experiments, they trained subjects on the classifications shown in Figure 4.6. In one problem (left panel), the stimuli can be partitioned by a plane constructed through the space, whereas for the other (right panel) this is not possible. Note that the same set of stimuli is used in the two cases. Medin and Schwanenflugel presented their subjects with photographs of people who differed in terms of hair length, hair colour, and type of smile, and instructed subjects that these were the only relevant dimensions. Subjects saw one of the faces on each trial and classified it, with corrective feedback, into one of two families. The results showed that while both category structures could be learned, the nonlinear classification was no harder to learn (in fact, slightly easier) than the linear one. In another study in which four-dimensional geometrical stimuli were used, the nonlinear classification was actually learned significantly faster than the linear one, a result that has also been obtained by Nakamura (1985).

Actually, this result is quite intuitive, because examination of Figure 4.6 shows that in the linear classification, each item's nearest neighbour is a member of the alternative category, which would be expected to impair learning. This is not true of the nonlinear classification: most items are equally similar to a member of their own category and to a member of the opposite category. At any rate, the important point for present purposes is that the broadly equivalent ease of learning of linear and nonlinear cat-

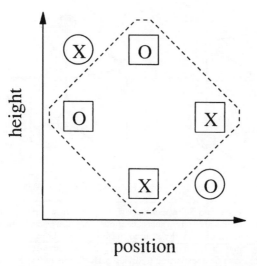

position

Figure 4.7. Category structure used in Kruschke's (1993) interference experiment. The stimuli were boxes with internal lines, and the height of the box and the position of the internal line varied as shown. In the first phase of the experiment, the stimuli in squares (inside the dotted line) were presented with category feedback, while the stimuli in circles (outside the dotted line) were classified with feedback only on every fifth presentation. In the second phase, the stimuli inside the dotted line were classified without feedback and those outside the dotted line were classified with feedback on every trial. The two categories are designated X and O. (From Kruschke, 1993, reprinted with permission.)

egories is hard to reconcile with the backpropagation model. Gluck (1991) has shown in simulations of Medin and Schwanenflugel's data that a standard backpropagation network predicts very much faster learning of the linear classification.

The second problem with standard backpropagation as a model of human associative learning comes from a rather different source. In the last chapter we saw that of the two major explanations of forgetting, unlearning and retrieval failure, the bulk of the evidence supports the latter. Indeed, we saw that there is actually little direct evidence of unlearning at all. To the extent that they can behave like instance memorisation systems, connectionist models find these results concerning forgetting entirely congenial. But whereas genuine unlearning seems to play a rather minor role in normal human forgetting, it appears that backpropagation networks are extremely prone to unlearning; indeed, they seem to suffer from an effect known as 'catastrophic interference', whereby target information is almost entirely overwritten or unlearned by later interfering information in a way quite uncharacteristic of human performance.

McCloskey and Cohen (1989) observed that a multilayer network using the backpropagation algorithm will perform extremely poorly at reprodu-

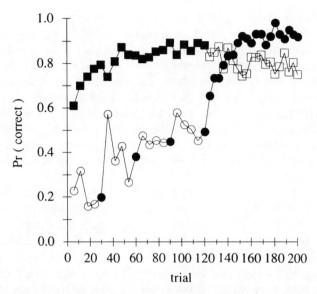

Figure 4.8. Results of Kruschke's (1993) interference experiment. The stimuli and categories were as shown in Figure 4.7. From trials 1–120, the stimuli inside the dotted line in Figure 4.7 were classified with category feedback on each trial (filled squares), while those outside the dotted line were classified without feedback on four out of five trials (open circles) and with feedback on the fifth trial (filled circles). From trials 121–200, feedback was now withheld from the stimuli inside the dotted line (open squares) but was presented on every trial for the stimuli outside the dotted line (filled circles). Filled symbols represent trials with feedback, open symbols represent trials without. The key result is that little unlearning of the category assignments occurred for stimuli inside the dotted line during the second phase. (From Kruschke, 1993, reprinted with permission.)

cing a set of associative A–B relations if it is taught some other information following these pairings. It is very difficult for such a network to maintain a record of a set of information in the face of some new information that has to be learned. In the extreme case, McCloskey and Cohen found that one set of input-output pairings (A–B) was entirely abolished in a backpropagation network by a new set of A–C pairs involving the same input cues.

Kruschke (1993) has conducted a more systematic demonstration of this effect, commencing with an experiment showing that humans are not especially prone to catastrophic forgetting. The stimuli used in the experiment, represented in Figure 4.7, were boxes with internal lines with the height of the box and the location of the internal line varying independently. In the first phase of the experiment, a stimulus was presented on each of 120 trials and the subject made one of two category responses. Corrective feedback was provided every time one of the stimuli inside the dotted line in Figure 4.7 (denoted by squares) was presented. For the two stimuli outside the dotted line (denoted by circles), feedback was only given on every fifth occasion. The data obtained by Kruschke are shown in Figure 4.8. As the filled squares in the figure show, performance to the stimuli inside the dotted line rapidly improved. For the other two stimuli, subjects initially classified them incorrectly: unsurprisingly, they were assigned to the categories to which they were most similar. However, with occasional feedback performance improved during the first phase. The zig-zag shape of the curve for the circle stimuli reflects the fact that after feedback was given (filled symbols), performance was boosted the next time the stimulus was presented, but that in the absence of feedback subjects soon reverted to classification based on similarity and hence made fewer correct decisions. By the end of the first stage (trial 120), performance on the circle stimuli was still only at about 50% correct.

In the second phase, the circle stimuli were now given explicit feedback on every trial whilst the square stimuli were given no category feedback. Subjects clearly did not suffer from catastrophic interference, since responding to the stimuli inside the dotted line did not deteriorate very much (open squares in Figure 4.8) in the face of new learning concerning the circle stimuli. Following McCloskey and Cohen's (1989) results, the critical question is whether a back propagation network will be able to reproduce the subjects' behaviour or whether it will show rapid forgetting of the square stimuli during the second stage.

To test this, Kruschke presented a multilayer network (using the back-propagation rule) such as that shown in Figure 4.5 with exactly the same sequence of events as his subjects had experienced. The network had two input units representing the two dimensions of variation of the stimuli, six hidden units, and two output units (one for each category), and Kruschke adjusted the learning rate to achieve the best fit possible. The results are shown in Figure 4.9. The network fails in a number of respects. First, feedback on the circle stimuli did not allow performance to improve at all during the first stage. This is because the learning that did occur on these trials was immediately interfered with by continued learning about the square stimuli. Secondly, and more importantly, the network was utterly incapable of reproducing the subjects' behaviour during the second stage of the experiment. Figure 4.9 shows that performance on the square stimuli fell to chance (50% correct) during the second stage, again reflecting catastrophic interference: learning about the circle stimuli completely undid learning about the square stimuli.

Why does the backpropagation learning algorithm lead to this inappropriate behaviour? The reason is that each of the hidden units in such a network is activated by a very high proportion of input patterns, which means that new patterns are very likely to lead to adjustments in the weights connected to a given hidden unit. Such weight changes, of course, almost inevitably entail unlearning. The large receptive fields of the hidden units mean that it is very difficult for a backpropagation network to isolate a particular item of knowledge and protect it from interference.

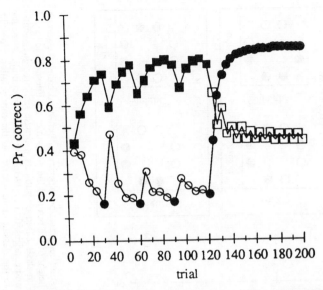

Figure 4.9. Simulation with a backpropagation network of Kruschke's (1993) interference experiment. The network consisted of two input units, six hidden units, and two output units, and was presented with the same trial types as subjects (see Figure 4.7). In the first phase, some stimuli were presented with feedback (filled symbols) and others without (open symbols), while in the second phase these assignments were reversed. During phase 2, complete unlearning of the phase 1 assignments occurred (open squares), but this did not occur with subjects (Fig. 4.8). (From Kruschke, 1993, reprinted with permission.)

The above results all suggest that there is a serious problem with the way in which hidden units in a backpropagation network represent stimuli. While there can be little doubt that internal representations are required in some situations, the way in which the generalised version of the delta rule changes weights, coupled with the logistic activation functions of the hidden units, means that it does not provide a good model of human behaviour.

Selective attention and learning

In Chapter 3 we examined some reasons why it is necessary to incorporate a selective attention process into models of associative learning. Recall that Nosofsky (1987) found that the context model was only able to account for data from classification experiments on the assumption that the psychological space in which the stimuli fell could be stretched or shrunk along its axes during the learning phase. For instance, when values on a certain dimension distinguished members of two categories, the space appeared to be stretched along this dimension and shrunk along nonpredictive dimensions. A typical learning task will therefore entail two processes, one

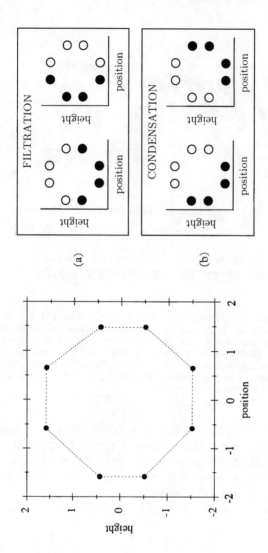

Figure 4.10. Stimuli used in Kruschke's (1993) filtration/condensation experiment. Left-hand panel: the stimuli (as before) were boxes with internal lines, and the height of the box and the position of the internal line varied as shown. These co-ordinates are from an MDS analysis based on subjects' pairwise similarity judgments. Right-hand panel: category partitions in filtration and condensation problems for this set of stimuli. In the filtration problems, only one dimension is relevant to the classification, whereas in the condensation problems both dimensions are relevant. (From Kruschke, 1993, reprinted with permission.)

whereby stimuli come to be associated with outcomes and one which alters the perceived inter-stimulus similarities.

Plainly, if our goal is to construct an adaptive network model of learning, we need to say something about how selective attention is to be dealt with. In order to see what is required, consider a quite simple result that seems to demand an attentional explanation of the same sort postulated by Nosofsky (1987). This is the classic filtration/condensation effect first studied by Posner (1964). Suppose subjects have to learn to classify stimuli varying on two dimensions; the stimuli might be red or green triangles or squares. In a filtration task, only one of the dimensions is relevant to the classification: the triangles are members of one category and the squares are members of the other category, regardless of colour. In a condensation task, both dimensions are relevant: one category consists of red triangles and green squares, the other of green triangles and red squares. Intuitively, the filtration task should be easier to learn because one of the dimensions can be filtered out and ignored, whereas in the condensation task information from both dimensions has to be condensed into a single classification decision.

Note that the filtration/condensation difference is not the same as the difference between linearly-separable and nonlinearly-separable categories, because linearly-separable problems can require either filtration or condensation. If red triangles and squares represent one category while green triangles and squares represent the other, then the problem requires filtration (of the shape dimension), whereas if red triangles, red squares, and green triangles represent one category and green squares the other then the problem requires condensation. But in both cases, the categories are linearly separable.

It is well-established that the intuition that people will find filtration tasks easier than condensation ones is correct. A particularly elegant example is provided in another study by Kruschke (1993). He used the same stimuli as described before, namely boxes with an interior line, in which the height of the box and the position of the line varied orthogonally. Kruschke used eight stimuli organised as shown in the left-hand panel of Figure 4.10, with the stimuli constructed from four levels of each dimension. In fact, the figure shows the co-ordinates of the stimuli obtained from a multidimensional scaling solution based on similarity judgments from a separate group of subjects.

Kruschke then presented subjects with 64 training trials in which four stimuli were assigned to one category and four to the other. Category feedback was provided on each trial. The right-hand panel of Figure 4.10 shows that in the filtration task, the category boundary was either a horizontal line separating the four upper stimuli from the lower ones or a vertical line separating the four stimuli on the right from those on the left. In the condensation task, the boundary was either the major or the minor diagonal. Figure 4.10 shows that in terms of inter-stimulus similarities, these classifications are all identical. Thus if learning simply requires associating stimuli

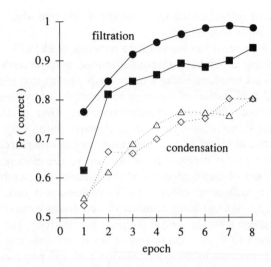

Figure 4.11. Results from Kruschke's (1993) filtration/condensation experiment. The figure shows the probability of a correct response for stimuli from the filtration and condensation classifications shown in the right-hand panel of Figure 4.10. The key result is that filtration problems are easier to learn than condensation ones. Filled circles = position-relevant filtration problem; filled squares = height-relevant filtration problem; open symbols are for the two condensation problems. (From Kruschke, 1993, reprinted with permission.)

with outcomes, no filtration/condensation difference would be expected. On the other hand, the presence of a difference would imply that, in addition to the basic stimulus→outcome learning process, there exists an attentional process which changes the perceived inter-stimulus similarities in such a way as to make the classification task easier to learn.

Figure 4.11 shows the results in terms of the probability of a correct classification in each of the tasks. The filtration classifications were significantly easier than the condensation ones. The position-relevant filtration task was also somewhat easier than the height-relevant one. How do these results relate to network models of learning? Most simple connectionist models find the filtration/condensation effect extremely difficult to explain, because other things being equal, they have no preference for learning a decision boundary in one orientation compared to another. As an illustration, Figure 4.12 shows the best fit that Kruschke was able to obtain using a simple backpropagation network consisting of two input units, eight hidden units, and two output units. The network is quite unable to capture the filtration/condensation difference.

The difficulties presented to connectionist models by the results discussed in this section and the preceding one lead us to two conclusions. First, it is necessary to incorporate some mechanism for selective attention that is independent of the basic process whereby the elements of the stimulus come to be associated with the outcome or category. Secondly, in order to avoid

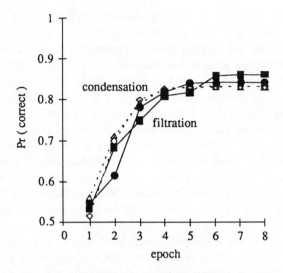

Figure 4.12. Simulation with a backpropagation network of Kruschke's (1993) filtration/condensation experiment. The network consisted of two input units, eight hidden units, and two output units, and was presented with the same trial types as subjects (see Figure 4.10). The network is unable to reproduce the filtration/condensation difference that appeared in the subjects' data (Figure 4.11). (From Kruschke, 1993, reprinted with permission.)

the problem of catastrophic interference, it is necessary to construct a network in which the hidden units are rather more constrained, in terms of the number of input patterns that they are activated by, than is the case in standard backpropagation. Catastrophic interference comes about because interpolated learning overwrites earlier learning, but this can be to some extent alleviated if the interpolated learning adjusts the weights of a different set of hidden units than that which encodes the original learning.

These two considerations have been taken into account in the construction of a model called ALCOVE (Kruschke, 1992; Nosofsky and Kruschke, 1992) which is a promising alternative to standard backpropagation. Briefly, hidden units in ALCOVE are strictly limited in the input patterns they respond to. In fact, a given unit is maximally activated by only one input stimulus; other stimuli activate it to a lesser degree that depends on how similar they are to the stimulus that yields the maximal response. Because of these hidden unit 'receptive fields', the ALCOVE model does not need to adjust weights connecting the input and hidden units. Instead, only the weights from the hidden to the output units are modifiable, but the way in which the hidden units represent input stimuli is still sufficient to allow nonlinear classifications to be learned.

Additionally, Kruschke incorporates into the model a selective attention mechanism whereby the activation of each input unit is multiplied by an attentional gain factor. Just as the error on the output units of the network drives weight changes (according to Equation 4.3), so Kruschke uses this error to adjust the amount of attention that is paid to the activation of a given input unit, and the end result is that the inputs that are most relevant to solving the classification acquire large attention strengths, while irrelevant inputs acquire small ones. As Kruschke (1993) has shown in an impressive simulation of the data shown in Figure 4.11, this allows the filtration/condensation difference to be very straightforwardly accommodated. In addition, because each hidden unit in the model is activated by relatively few stimuli, catastrophic interference does not occur. Knowledge stored across the weights connected to one set of hidden units is relatively unaffected by subsequent changes in the weights of a different set of hidden units. Accordingly, Kruschke (1993) has shown that the model can fit very closely the data of Figure 4.8 which were so problematic for backpropagation.

This brief discussion of the ALCOVE model illustrates how connectionist models are currently being extended to allow them to reproduce some of the more subtle aspects of human learning. At present, it is too soon to evaluate the ALCOVE model in great detail, but there is no doubt that it can fit the data from a very large number of experiments (Nosofsky and Kruschke, 1992). This is a fast-moving area of research and promises to yield further insights into the mechanisms of associative learning.

Configural cues

One interesting aspect of Kruschke's ALCOVE model is that it represents stimuli in a configural rather than an elemental manner. In a single-layer network, it is assumed that the elements or features that constitute the stimulus are independently associated with the category or outcome, whereas in ALCOVE it is not the elements themselves but rather specific configurations of elements that form associations. Each hidden unit in the network represents a specific stimulus configuration.

We have already been given abundant reasons to reject single-layer networks, but the configural–elemental dimension provides a further reason in the form of direct evidence that people often learn about configurations rather than about elements. In the present section we will examine some of the evidence for the sort of configural representation assumed in Kruschke's model. As a first example, let us consider an experiment that Wilson (1993) conducted which was modelled on an earlier animal conditioning study of his. Subjects saw intermixed $A \rightarrow O_1$, $AB \rightarrow$ no O, and $BC \rightarrow O_1$ trials together with $D \rightarrow$ no O, $DE \rightarrow O_2$, and $EF \rightarrow O_2$ trials. After the learning phase, Wilson presented subjects with two new test trials consisting of the compounds ABC and DEF, and asked subjects to predict the outcome most likely to occur on each trial type.

Wilson reasoned that on an elemental theory, ABC should be associated with outcome O_1 to a greater degree than DEF is associated with outcome O_2. The delta rule predicts that after training with $A \rightarrow O_1$, $AB \rightarrow$ no O, and $BC \rightarrow O_1$ trials, A should have an asymptotic weight of 1.0, B of -1.0, and C of 2.0, while after $D \rightarrow$ no O, $DE \rightarrow O_2$, and $EF \rightarrow O_2$ trials, D should have a weight of 0.0, E of 1.0, and F of 0.0. When these are added together, the net result is a combined weight of 2.0 for ABC and 1.0 for DEF. But in contrast to these predictions, Wilson's subjects were significantly more likely to predict outcome O_2 on a DEF test trial than outcome O_1 on an ABC trial.

Why should this be? Suppose that subjects learn not about the individual associations between elements and outcomes, but rather between whole configurations and outcomes as in the ALCOVE model. Thus, they learn that AB is associated with no outcome and BC with outcome O_1, but they do not directly learn anything about element B. On this account, responding to a novel combination such as ABC must be determined not by the associative strength of its elements but rather by its similarity to previously-seen configurations. Thus ABC is highly similar to AB and BC, but only one of these has been paired with the outcome. In contrast, DEF is highly similar to DE and EF, and both of these have been associated with the outcome. Thus Wilson's results are compatible with the idea that the outcome may be associated with the whole configuration rather than with the separate elements of which it is made.

Another illustration of the same effect comes from a series of experiments by Williams (1995) on the learning of negative contingencies. Williams used the fictional stock market procedure of Chapman and Robbins (1990) in which the cues are stocks and the outcome is a rise in the overall level of the stock market. In his studies, subjects were exposed to intermixed $A \rightarrow$ outcome and $AB \rightarrow$ no outcome trials designed to imbue cue B with negative associative strength. Together with these training trials, subjects also saw trials with another cue that was also paired with the outcome ($C \rightarrow$ outcome). By the end of training, subjects were predicting the outcome reliably less often on the AB than on the A trials, indicating that they had learned about the different consequences of these trial types. Then in the test stage cue B was presented either alone or together with cue C and Williams recorded whether subjects predicted the outcome or not.

On the elemental assumption that the cues were directly associated with the outcome, the initial $A \rightarrow$ outcome, $AB \rightarrow$ no outcome trials should have left cue B with a negative association, and we would obviously predict that subjects will judge the outcome to be less likely on the B test trial than on a trial with a completely novel cue D which should have zero associative strength for the outcome. In fact, in this experiment Williams found no difference between responding to B and D, contradicting the idea that B had a negative weight. Moreover, there was no difference in responding on BC compared to CD test trials, where again a difference would be expected on an elemental analysis, since D would be predicted to be neutral while B is predicted to have a negative weight. What seems to have happened instead is that the initial discrimination taught subjects that cue A was paired with the outcome and that the configural cue AB was paired with no outcome. Subjects did not learn anything directly about the relationship between B and the outcome.

Williams went on in a further study to show that an elemental association could be formed under slightly different training procedures. Here, subjects received A→O, AB→no O, C→O, and B→no O trials with B now being presented alone on some trials without the outcome. This direct exposure to the relationship between B and the outcome was sufficient to generate an elemental negative association, in that on test trials subjects predicted the outcome reliably less often with cue B than with a novel cue D. In sum, it appears that a configural association will control performance unless the subject has seen the elements in isolation, in which case the tendency to form a configural representation seems to be attenuated.

Unfortunately, this picture is complicated by some evidence that subjects appear to be able to adopt either an elemental or a configural approach to a given task depending merely on prior experience. Williams, Sagness and McPhee (1994) repeated Chapman and Robbins' (1990) blocking experiment (see Table 2.5) but pretrained various groups of subjects in different ways. In one condition the pretraining was designed to foster an 'elemental' strategy whereby subjects would to some degree analyse each cue separately. Specifically, the pretraining phase involved exposure to intermixed X→O and XY→O trials, and although the XY configuration is paired with the outcome, there is explicit information suggesting that it is the X element of the configuration that is the important one. In a second condition, pretraining was designed to foster a 'configural' strategy. Here, subjects received XY→O, X→no O, and Y→no O trials; clearly, in this case it is the XY configuration rather than either of its elements that predicts the outcome.

After this pretraining phase, subjects then went on to the main phase of the experiment in which they received a standard blocking problem using a new set of cues. In the first stage A→O and B→no O trials were presented, followed in the second stage by AC→O and BD→O trials. What would we expect to happen in this situation when subjects finally rate cues C and D? Given that the main phase is simply a replication of Chapman and Robbins' (1990) experiment, we should expect to see that cue C is blocked and receives lower ratings than cue D, and this is exactly what happened for subjects who received the elemental pretraining. However, for those given the configural pretraining, no blocking was observed, and instead C and D received equal ratings.

To explain this intriguing result, Williams *et al.* (1994) argued that when subjects saw the AC and BD trials, they could either treat these compounds as being composed of separable elements or as constituting configurations. In the former case, analysis will reveal that cue A is more likely than cue C to be the element most predictive of the outcome and that cue D is more likely than cue B to be predictive, and hence blocking will be observed. In contrast, if the subjects are inclined to treat the AC and BD configurations as being relatively unrelated to the A and B elements seen in the earlier stage, then they should be treat them equally, since each is paired to the same extent with the outcome. In this case, no blocking would be expected.

The success of Williams *et al.*'s pretraining manipulations to bias the way subjects rated the cues in the blocking part of the experiment suggests that this strategic theory is correct. Of course, we can account for the elemental strategy by assuming a network with independent but direct connections from the cues to the outcome. At the same time, we can account for the configural strategy in terms of a network in which the elements feed into a configural unit as in the ALCOVE model, with this single configural unit being connected to the outcome. What we are at present unable to explain is how subjects seem to be able to switch from one mechanism to another simply on the basis of recent experience. The interaction of elemental and configural approaches to associative learning will clearly become a major focus of future research.

Perceptual learning

Selective attention operates to modify inter-stimulus similarities in situations where some features or dimensions are more predictive of a category than others, and in our discussion of the context model in the last chapter we considered some evidence from a study by McLaren *et al.* (1994) that stimuli become more discriminable from one another during the course of a category learning experiment. But it has also been recognised at least since the work of William James at the end of the last century that exposure to a stimulus, *in the absence of any overt consequences*, tends to alter the ease with which it is discriminated from other similar stimuli. James' famous example concerns the novice who starts out being unable to distinguish claret from burgundy, but eventually, simply as a result of extended exposure to these wines, comes to find them highly distinct. Another familiar example is that we tend to be better able to discriminate the faces of people from our own racial and ethnic group than those of other groups. The process whereby stimuli become more discriminable as their familiarity increases is called *perceptual learning*. Because it can occur in the absence of overt outcomes, perceptual learning cannot be explained by Kruschke's ALCOVE model even though the model incorporates a selective attention mechanism.

In the laboratory this sort of perceptual learning effect is readily demonstrated. For instance, in a famous experiment Gibson and Walk (1956) gave laboratory rats prolonged exposure to circles and triangles hanging on the walls of their cages, and found that the animals were subsequently better able to learn a discrimination between these stimuli, where they were paired with different outcomes, than were animals not preexposed to the stimuli. The circles and triangles became more distinct from one another even in the absence of differential outcomes in the pre-exposure phase. Another classic study, by Attneave (1957), demonstrated a similar result with geometrical patterns in humans. In Attneave's experiment, subjects were exposed in the first stage of the experiment to the prototype of a category of patterns. In the second stage, subjects learned identification responses to each of a number of stimuli that were generated from the prototype. Attneave obtained a perceptual learning effect in that identification learning was faster for subjects pre-exposed to the prototype than for those who had not seen the prototype. This suggests that during the pre-exposure phase, subjects learned something that made exemplars generated from the prototype more discriminable.

How can we begin to understand the processes responsible for perceptual learning? As it happens, we have already considered a theoretical approach that might be relevant. In an artificial grammar learning experiment, the subject is initially exposed to a set of strings generated from a grammar. Since these strings are presented without any overt feedback, we would naturally expect perceptual learning to occur with subjects coming to find the strings progressively less similar to one another. We might expect a similar outcome in Whittlesea's (1987) experiments in which subjects saw letter strings generated from prototypes such as FURIG.

Although we have no direct evidence of perceptual learning in either of these cases, what we do have is a theoretical analysis of how learning proceeds during exposure to these strings, and so we can ask whether this account is able to predict perceptual learning. We considered at length the connectionist model Dienes (1992) proposed for artificial grammar learning, and also the model McClelland and Rumelhart (1985) proposed to account for Whittlesea's data. In each case, an autoassociative network which learns associations amongst the elements making up each stimulus is capable of explaining a good deal of the relevant data we considered.

Let us imagine that we have two stimuli A and B each made up of a large number of elements. Naturally, these stimuli will share some common elements, but each will also possess unique elements. If we designate the common elements by x and the unique elements by a and b, then $A=a+x$ and $B=b+x$. Early on, the patterns of activation that these stimuli elicit in an autoassociative network (governed by Equations 4.1–4.3) will be quite weak since the weights connecting the elements will be small and relatively little activation will pass between the units in the network. As training continues, the activation patterns for the two stimuli will begin to develop. Unfortunately, however, there is no reason to believe that the activation patterns will diverge, and in fact it is quite possible that the opposite will happen. Since the features the network sees most often are the ones that are common to the two stimuli, the interconnections between these common features will gain most strength. This is, after all, the process that allows networks to show prototype enhancement effects and to show generic learning of the sort demonstrated by Watkins and Kerkar (1985). After prolonged training, the chances are that the patterns of activation in the network will be quite similar and will be dominated by the contribution of the common elements. Thus the gradual differentiation of the activation patterns that occurs in perceptual learning will not be achieved.

A more promising solution can be seen by considering the analysis of perceptual learning proposed by McLaren *et al.* (1989). Although these authors suggested that a number of possible mechanisms may be involved, they focused in particular on one process that has no counterpart in a standard autoassociator. This mechanism has its origins in a paradoxical pair of observations. We have already discussed how pre-exposure to a stimulus makes it more discriminable from other stimuli and therefore enhances later learning, but as we saw in Chapter 2, in other circumstances it is known that pre-exposure will *retard* rather than enhance learning. Abundant evidence from latent inhibition experiments such as that of Lipp *et al.* (1991) suggests that when a stimulus is presented alone, later learning concerning that stimulus will be adversely affected. Thus, exposure to a stimulus not only renders it less able to enter into new associations but also makes it more discriminable from other similar stimuli. How does this come about?

Although these processes seem contradictory, McLaren *et al.* have suggested that they have a common cause. These authors propose that when a stimulus is pre-exposed, attention to its elements is steadily reduced. Although connections will be formed amongst the elements of the stimulus (as in an autoassociative network), if the stimulus is later paired with an outcome as in Lipp *et al.*'s experiment, learning will be retarded since little attention is focused on the stimulus. Hence latent inhibition is explained. Turning to perceptual learning, McLaren *et al.* note that presentation of A ($= a+x$) and B ($= b+x$) will lead to greater latent inhibition of the common than of the unique elements. Because the common elements are seen twice as often as the unique ones, attention will decline to the common elements far faster than to the unique elements, which in turn means that the representations of the stimuli will tend to become differentiated. When A is now paired with one outcome and B with another, it is the a and b elements that acquire virtually all of the associative strength. If the stimuli had not been pre-exposed, then one outcome would be associated with $a+x$ and the other with $b+x$, a situation that would clearly lead to greater generalisation and hence slower learning.

In a nutshell, the conclusion is that some mechanism is required for reducing the amount of attention paid to the elements of a stimulus presented in the absence of significant consequences. While a number of specific suggestions have been proposed (see McLaren *et al.*, 1989), the simplest procedure is to adapt the delta rule equations such that the learning rate parameter α is steadily reduced for a stimulus that occurs with no significant consequences. We will not go into the details of such mechanisms, but it makes good sense that evidence that an event is of no predictive value should encourage the learning system to pay less attention to that event.

Retrospective revaluation

This chapter should have made it clear that the enormous amount of attention being paid to connectionist models of learning is well-justified. These models offer the hope of explaining a huge range of associative learning phenomena. At the same time, no-one would deny that major challenges lie ahead for the connectionist research programme, and I end the chapter by considering one particular line of evidence that seems to expose a major problem with the delta rule. This evidence comes from demonstrations of the retrospective revaluation of cue strength.

We have considered at length Chapman's (1991) experiment demonstrating an effect of trial order on ratings of negative contingencies (Chapter 2, Table 2.8). Remember that subjects gave less negative ratings to cue D after observing CD→no O followed by C→O trials than they gave to cue B after A→O followed by AB→no O trials. As we saw earlier in this chapter, the delta rule can account for the fact that judgments differed in these two conditions, but actually the model makes a further prediction which was not borne out by the data: it predicts that judgments for cue D should have been close to zero, whereas in fact they were quite strongly negative. According to adaptive connectionist models, the CD→no O, C→O procedure should not have been able to endow cue D with any negative strength at all, yet that is apparently what happened. D should have zero associative strength after the first stage and this should be unaffected by the C→O trials.

Of course, just because D received a negative rating does not mean that it had a genuinely negative associative strength: Chapman's experiment does not include a control condition providing a neutral stimulus against which cue D can be compared. Therefore Chapman conducted a further experiment in which subjects now received intermixed AB→no O and CD→no O trials in the first stage followed by C→O trials in the second stage. Here, we can compare ratings given to cues D and B at the end of the second stage, and any difference between them will be evidence of negative associative strength. In this design, cue B represents a control stimulus against which D can be assessed. The results shown in Figure 4.13 reveal that D is genuinely more negative than B. Although the effect is small, the C→O trials do seem to have made the weight connecting cue D and outcome O more negative.

The reason that connectionist systems using learning algorithms like the delta rule have difficulty explaining cue D's associative rating is that a cue's

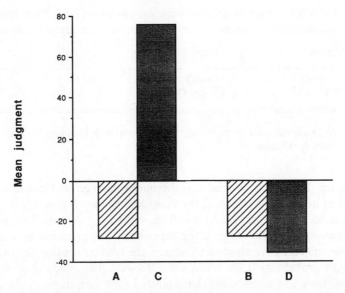

Figure 4.13. Retrospective revaluation effects. Subjects were presented with AB→no O and CD→no O trials in the first stage followed by C→O trials in the second stage, where O is a fictitious disease and no O is no disease. Then they judged the contingency between each of the four stimuli and outcome O. The key result is that cue D is rated more negatively than cue B. (After Chapman, 1991.)

associative strength can only be changed if it is present on a trial. Recall that Equation 4.3 states that the weight change Δw_{io} is given by:

$$\Delta w_{io} = \alpha\, a_i\, d_o\,.$$

If the activation a_i of the input unit representing the cue is zero – that is, if the cue is absent – then no weight change should occur, since the error term d_o is multiplied by a_i to determine the weight change. In the second stage of Chapman's experiment, cue D is absent, so according to this equation there should be no change in its weight. Yet when cue D's associative strength is assessed, it is found to have been retrospectively revalued.

Another example of retrospective revaluation that lies outside the scope of current connectionist models is known as 'backward blocking'. The term refers to the fact that the blocking effect that we discussed in Chapter 2 occurs when the two stages of the experiment are reversed. Two demonstrations should serve to illustrate the effect, one within the domain of action–outcome learning, the other involving cue–outcome learning. In the first (Shanks, 1985), subjects were allowed to perform an action in a certain context and judged the extent to which the action caused an outcome. The action involved pressing a key on a computer keyboard which fired a shell at a tank passing across the screen. The outcome consisted of the tank

Table 4.4. *Design of the experiment by Chapman (1991, Experiment 1)*

Stage 1	Stage 2	Test trials
AB→O	A→O	B, D
CD→O	C→no O	

A–D are the cues (symptoms), O is the outcome (disease), and no O indicates no disease.

blowing up, and on each trial an alternative background cause was the presence of an invisible minefield the tanks had to pass through.

For the control group P(O/A) was 0.75, while P(O/–A) was 0.50. For one experimental group, this stage of the experiment was preceded by a stage in which the subjects were allowed to witness the context causing the outcome, again with probability 0.50. When these subjects then made their causality estimates, their judgments were significantly lower than those of the control group who had not witnessed the prior stage, thus yielding a forward blocking effect. Essentially the result confirms that if the background B is paired initially with the outcome (B→O), subsequent learning of the action-outcome relationship is impaired when the action and background co-occur (AB→O), an effect we have seen is easily explained by the delta rule. The important condition in the experiment was identical to this forward blocking condition except that the two stages were reversed: subjects witnessed the action-outcome contingency before observing the background-outcome contingency, and the result was that blocking occurred in this group just as in the forward group. As in the forward case, being given extended experience of the background causing the outcome allowed the subjects to make the apparent inference that the action was not the primary cause of the outcome.

There is little doubt of the robustness of the backward blocking effect in humans. Although she used a different learning task, Chapman (1991) also obtained the effect. In her experiment, illustrated in Table 4.4, subjects judged relationships between symptoms and a fictitious disease. In the first stage, different patients had either symptoms A and B or symptoms C and D, and all had the disease. In the second stage, further patients had symptom A and the disease or symptom C and no disease. After witnessing 12 trials of each trial type in each stage, judgments were made of the symptom–disease relationship for each symptom. After the first stage subjects rated A, B, C, and D about equally, as expected. The key results concern ratings at the end of the second stage. Figure 4.14 illustrates that after the second stage higher ratings were given, as expected, to A than to C, but also that backward blocking occurred, since cue D was given a higher rating than cue B despite the fact that these cues were rated equally after the first stage and had not been presented during the second stage. Their status as predictors of the disease had been retrospectively revalued.

From a psychological point of view backward blocking is not a surprising finding because if one learns that a pair of events is predictive of an outcome, and that one of those events is sufficient on its own, then it makes sense to infer that the other event was not responsible. Of course, this inference is not *logically* necessary, since each cue might be sufficient on its own, but nevertheless the result does not seem especially unusual. Perhaps it could be accommodated within a connectionist model simply by removing the term a_i from Equation 4.3 and allowing weights of absent cues to be changed? The problem with this solution is that the weight change would be in the wrong direction. With AB→outcome, B→outcome trials, there will be an increase during the second stage in the weight connecting cue B with the outcome, and we want this to lead to a *decrease* in the cue A–outcome weight. Removing the term a_i from Equation 4.3 would fail to achieve this.

A more promising alternative has been suggested by Van Hamme and Wasserman (1994). They also suggested that the weights of absent cues be allowed to change, but in this case by setting a_i to a negative value in Equation 4.3 rather than to zero. This produces a change in the opposite direction for an absent cue compared to a present one, just as we require. Of course, in general new knowledge does not typically lead to the unlearning of older knowledge. To work, Van Hamme and Wasserman's solution would have to restrict weight changes only to absent cues that in the past have co-occurred with the cue that is present; in Chapman's experiment, learning about the relationship between cue A and the outcome in the second stage should only lead to weight changes for cue B and not for the millions of other cues that are simultaneously absent.

Plainly, more work is needed to elaborate this solution and determine whether it is a satisfactory way to account for backward blocking, but at present it does appear that a fundamental assumption of associationist models – namely that new learning about a given cue can only occur when the cue is present – may in some circumstances be violated.

Relationship to prototype and instance theories

The discussion of connectionist models in this chapter will probably have given the impression – contrary to what I have been arguing – that they are quite different from, and indeed inconsistent with, the sorts of theory considered in Chapter 3. After all, it has been possible to generate clear differential predictions from some of these theories. For instance, nonlinear classifications can be learned by connectionist models with hidden units but cannot be learned on the basis of prototype abstraction, and connectionist and instance theories predict different outcomes for Gluck and Bower's (1988) AB→1, AB→2, A→1 design.

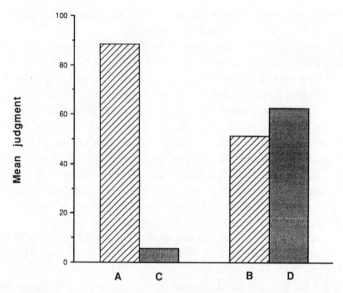

Figure 4.14. Backward blocking. Subjects were presented with AB→O and CD→O trials in the first stage, followed by A→O and C→no O trials in the second stage, where O is a fictitious disease and no O is no disease (see Table 4.4) The figure shows the ratings of A, B, C, and D at the end of the second stage, and the critical result (backward blocking) is the lower rating of B than D. (After Chapman, 1991.)

As might be expected, there has been an extended debate on the issue of whether connectionist models are inconsistent or not with other theories such as ones based on prototype abstraction or instance memorisation. What exactly is the relationship between them? In the original exposition of their connectionist model of learning, McClelland and Rumelhart (1985) argued forcefully that the model was not simply an implementation of a higher-level theory in distributed processing hardware. They appealed, for instance, to the fact that connectionist models of the sort considered in this chapter only learn when an error signal is present representing a discrepancy between what is expected and what occurs. Instances, in contrast, are memorised automatically regardless of any expectancy the subject might have, so perhaps some way could be found to test which of these views is correct? By citing such differences, McClelland and Rumelhart hoped to distinguish between the theories in terms of psychologically-important phenomena.

In response to McClelland and Rumelhart, Broadbent (1985) argued that rather than being an alternative to instance or prototype theories, the connectionist approach is at a different level of explanation in that it provides a detailed mechanism whereby the basic processes of those alternative theories are performed. Thus to say that associative knowledge is encoded in a set of weights in a distributed network is not necessarily to deny that, at a different level of description, knowledge is encoded as a set of memorised instances or as a prototype. A connectionist model, according to Broadbent, is simply a description of how the instances or prototype are stored. It should come as no surprise that I favour Broadbent's rather than McClelland and Rumelhart's position in this debate. Prototype and instance theories do not attempt to provide a specific mechanism whereby the precise computational processes involved in associative learning are described, and thus it is mistaken to challenge them with data related to the details of such a mechanism. Rather, these theories attempt to specify at the *informational* level what is required of a model of human learning. The answer to that question is that the principal requirement is for the ability to memorise training exemplars. The fact that connectionist models can account for such a broad range of data in no way challenges that conclusion.

Summary

The re-emergence of associationist theories of learning over the last two decades has been dramatic, and is mainly due to the development of error-correcting learning algorithms such as the delta rule. This rule can be shown to yield associative weights which at asymptote are equivalent to conditional contingencies calculated by the ΔP rule. Current connectionist models have been successful in accounting for a range of basic phenomena such as the effect of contingency on associative learning, as well as more complex effects such as enhanced responding to an unseen prototype pattern and partial memory for the training items. Single-layer networks are unable to learn nonlinear classifications, though, and so it is necessary to introduce internal units which change the way in which the stimuli are represented.

The best-known such procedure uses a generalised version of the delta rule called the backpropagation algorithm, and this allows weights both between the input and hidden units and between the hidden and output units to be adjusted. However, standard backpropagation networks suffer from at least two major problems. First, acquired knowledge suffers catastrophic interference from later learning, and secondly, there is no mechanism for selective attention. These can be remedied by changing the nature of the hidden units, as in Kruschke's ALCOVE model, such that they have much smaller receptive fields and are therefore able to avoid extensive interference from interpolated learning, and by introducing a mechanism whereby the activation of an input unit is adjusted to the extent that the unit is responsible for error on the output units. Finally, we have seen one phenomenon, retrospective revaluation, that questions the basic assumption of these models that the weight from a given cue can only be changed on trials when it is present.

5   Rule induction

The picture of concept learning that emerges from the previous chapters is of a rather passive process in which instances are encoded in memory as a result of weight adjustments in an adaptive network system. This is passive in the sense that so long as the subject attends to the stimuli, the hypothesised processes operate automatically on the incoming information. But it has commonly been argued that in some circumstances a different, active process can operate whereby a person considers various hypotheses concerning relationships between events, modifies or rejects inadequate hypotheses, and in short tries to induce a rule describing the relationship between stimuli and outcomes. In this chapter we consider the evidence that the account of associative learning discussed in the previous chapters is incomplete and needs to be supplemented by an additional and possibly independent rule-learning mechanism.

Before considering the evidence and nature of this rule-learning process, it is necessary first to consider what exactly we mean by a 'rule'. This concept has, to put it mildly, been a source of some debate and confusion amongst psychologists and philosophers. On the surface, the definition of a rule seems unproblematic: we simply say that a rule is a principle that specifies definitively whether an object or event is of a particular sort or not. For instance, if an object has four sides of equal length lying in a plane and with right-angles between them, then it is a square. Any object conforming to this principle is a square, and any object that violates the principle is not a square. Note that it makes no difference if the violation of the rule is major or minor: even if a quadrilateral figure has two internal angles of 89° and two of 91°, it is still not truly a square. Many other examples of rules can be found in the legal world. For example, in English law an act of theft is said to have occurred if a person dishonestly appropriates property belonging to someone else with the intention of permanently depriving them of it. Again, this is a rule in the sense that an act either conforms to the definition or it does not. No matter how 'theft-like' an act is, if the definition is not fulfilled (say because there was no intention permanently to deprive) then it does not count as theft.

Of course, rules such as these are objective, public entities, while what psychologists are interested in are mental representations. Anyone can inspect public rules such as laws because they are written down, but our access to mental rules, if they exist, is rather more indirect, which is why it is controversial as to exactly what counts as a mental rule. Our only evi-

152

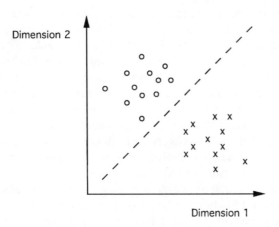

Figure 5.1. A set of hypothetical stimuli varying on two dimensions. If subjects respond on the basis of a test item's similarity to the category instances or prototypes, then classifications will be graded: test items close to the category boundary (the dotted line) will be classified less accurately and rapidly than those closer to the training stimuli of one of the categories. On the other hand, if subjects respond on the basis of a rule, then all stimuli falling on one side of the category boundary should be classified identically. O = category 1, X = category 2.

dence concerning their existence comes from observable behaviour. Psychologists and philosophers have been divided on what sort of behaviour is characteristic of rule-based knowledge, but in this chapter I shall follow Herrnstein (1990) and Smith, Langston and Nisbett (1992) in adopting the instrumental definition that behaviour is based on a rule if no difference is observable between performance to trained (old) and untrained (new) stimuli that fall into the same category.

To see why this definition is adopted, we must first remind ourselves of what the characteristics are of non-rule-based behaviour. Let us suppose that subjects in some category learning task improve their classification performance in the study phase either by memorising the training stimuli just in the way the context model proposes, or by abstracting the underlying prototype. Thus if the stimuli fall into two categories, as shown in Figure 5.1, subjects respond 'category 1' if a stimulus is more similar to the exemplars or prototype of category 1 than to the exemplars or prototype of category 2. Clearly, on this account responding to test stimuli is going to be graded: some new test stimuli will be highly similar to trained stimuli from one category, and hence will evoke rapid and accurate responses, whereas others will be more equally similar to training stimuli from the two categories and hence will be classified more slowly and less accurately. Of course, in Chapter 3 we saw that exactly such behaviour may be observed: for example, in Rosch *et al.*'s (1976) experiment (see Figure 3.3), classification times increased in an orderly manner as test items got further from the prototype.

By contrast to the behaviour expected if subjects abstract a prototype or memorise the training instances, we would expect to observe none of these differences if subjects learn and respond on the basis of a rule. Figure 5.1 shows a hypothetical boundary dividing the two categories. If this boundary perfectly divides members of one category from those of the other, then it constitutes a rule for classifying the stimuli: a stimulus on one side of the boundary is in category 1, one on the other side is in category 2. We would be strongly motivated to conclude that subjects have learned and are responding on the basis of this rule if the probability and latency of making a correct response is the same for all stimuli (whether old or new) falling on one side of the boundary, for such a result would suggest that the subject is merely analysing the stimulus to decide on which side of the boundary it falls and is not concerned in the least to compare it to previously-seen stimuli.

Thus suppose the stimuli in Figure 5.1 are rectangles varying in width and height, and suppose the two categories are defined by a rule that unequivocally assigns a stimulus to category 1 if its width is greater than its height and to category 2 if its height is greater than its width. If subjects are able to learn this rule from exposure to some training examples, and if they respond according to the rule, then when they make a classification decision they should merely be interested in whether the stimulus is wider or not than it is high; its similarity to training items should be immaterial. And if responding is based on a decision as to whether the stimulus is wider than it is high or not, that decision (ignoring what happens when width and height are perceptually difficult to discriminate) should be performed equally rapidly and accurately for all stimuli, regardless of how similar they are to stimuli seen in the training phase.

Of course, not all categories can be accurately described by an objective rule or boundary such as that shown in Figure 5.1. For instance, the random dot stimuli used in many laboratory experiments are generated by adding noise to each of two or three prototype patterns. Unless the prototypes are highly dissimilar or the amount of added noise is quite small, it is always a possibility that a given pattern could have been generated from more than one category. Unless there is only one correct response for each stimulus, it cannot be said that there is an objective rule for classifying the stimuli. But this does not mean that subjects do not still try to learn a rule by imposing a rule-based classification on the stimulus set: they may incorrectly come to believe that there does exist a classification rule.

The notion that rule-based learning is characterised by an absence of any detectable difference between performance to trained and untrained stimuli is intimately connected to the idea that people are able to form abstractions that go beyond the specific items they experience. Equivalent responding to new and old stimuli implies that an induction has been formed which governs responding to all stimuli. As we shall see below, it is relatively straightforward to obtain evidence for the formation of abstract representations capable of playing a role in associative learning.

Distinguishing rule- and instance-based behaviour

The reader may at this point be thinking that the distinction between rule- and instance-based behaviour is a very subtle one which is likely to prove extremely difficult to investigate empirically. Thus it is probably worthwhile considering a relatively simple experiment which illustrates fairly clearly how the predictions of the two theories may differ. A study by Perruchet (1994) provides just such an example. As it happens, the data from this experiment can be interpreted entirely in terms of instance- rather than rule-based responding. Nevertheless, it is useful to look at this experiment prior to considering others more successful at revealing rule-based knowledge, because the predictions of the rule-based account are particularly clear.

Perruchet's experiment is in fact a replication – with a minor change – of an earlier study by Kushner, Cleeremans and Reber (1991). These authors had asked subjects to make predictions about the location of a stimulus on a computer screen. This stimulus, a small square, could appear in one of three locations (A, B, or C) at the vertices of an imaginary triangle on the screen. Subjects observed the square moving in rapid succession from one location to another and appearing in a total of five locations. Then, the subject pressed a button to indicate where he or she thought it would appear next, and the stimulus moved to its next location. After a pause, the subject witnessed the stimulus move between five locations again, made another prediction, and this sequence repeated many times.

The position of the stimulus on a prediction trial was determined by a set of rules and depended on where the stimulus had been on the five preceding trials. In fact, its location on trials 1, 3, and 5 was irrelevant, but if it appeared in the same location on trials 2 and 4 then it appeared in location A on the prediction trial, if its movement from trial 2 to trial 4 was clockwise then it appeared in location B, and if its movement from trial 2 to trial 4 was anticlockwise then it appeared in location C. These rules are shown in Table 5.1. Clearly, these are quite simple rules, but they are embedded within a difficult task where three of the stimulus locations are irrelevant. Despite this, subjects were able to improve their accuracy when trained with over 4000 prediction trials spread across six days, although performance never reached a very high level, increasing from a chance value of 33% correct predictions to a final level of about 45% correct.

There are two ways of describing what the subjects learned. One possibility is that they acquired some knowledge about the underlying rules governing target location. Of course, since performance came nowhere near what would be achieved (100% correct predictions) if all three rules were perfectly learned, we would have to assume that most subjects were only learn-

Table 5.1. *Target locations and rules in Kushner et al.'s (1991) experiment*

Second event	Fourth event	Sixth event	Rule
A	A	A	
B	B	A	Same position
*C	C	A	
A	C	B	
C	B	B	Clockwise
*B	A	B	
B	C	C	
C	A	C	Anticlockwise
*A	B	C	

Location A is the top vertex of the triangle, B is the lower left and C the lower right vertex. Trial 6 is the prediction trial. Stimulus locations on trials 1, 3, and 5 were irrelevant to location on trial 6. *Sequences omitted from the training stage in Perruchet's (1994) experiment.

ing perhaps one or possibly two of the rules, or that they were applying their knowledge inconsistently. Nevertheless, it is perfectly possible that partial knowledge of the rules is what explains the observed improvements in performance. On the other hand, an alternative explanation is that subjects may merely have memorised some or all of the sequences and that each prediction was determined by the similarity of that sequence to previously-seen ones. Thus, if the subject saw the sequence ABCBA earlier in the experiment and learned that for this sequence the correct prediction was location A, then when a later similar sequence such as ABCBB was encountered, a prediction of location A may again have been made. Naturally, the evidence we examined in Chapter 3 for this sort of instance-based behaviour should encourage us to take this view very seriously.

Kushner *et al.*'s results can thus be interpreted in either of these ways, and Perruchet (1994) set out in his study to try to distinguish between them. How can this be achieved? The variation that Perruchet introduced was to omit some patterns from the training phase of the experiment and present them instead in a later transfer phase, and the reason for doing this is that the two theories make different predictions concerning performance on these test stimuli. In the training stage subjects saw two instances of the 'same position' rule but not the third, two instances of the 'clockwise' rule but not the third, and two out of three instances of the 'anticlockwise' rule. This procedural change did not affect the basic data from Perruchet's learning phase, which is that (as in Kushner *et al.*'s experiment) subjects were

Table 5.2. *Predictions of rule- and instance-based models and results of Perruchet's (1994) experiment*

Second event	Fourth event	Rule model	Instance model	Proportion of predictions		
				A	B	C
C	C	A	B or C	0.20	0.38	0.42
B	A	B	A or C	0.49	0.21	0.29
A	B	C	A or B	0.43	0.31	0.26

able to increase their prediction accuracy. In this experiment, performance increased from about 33% to 41% correct.

What are the predictions for the transfer phase? Table 5.2 shows the relevant sequences and the predictions of the two theories. If subjects are learning and responding on the basis of knowledge of the rules, they should behave in accordance with the rules given in Table 5.1. Thus for a sequence in which the stimulus appears in location C on trials 2 and 4, subjects should select location A as their response, regardless of stimulus location on trials 1, 3, and 5. Despite the fact that this is a new, untrained sequence, it should be responded to just as accurately as the old, trained sequences which define the rule. Of course, this prediction has to be qualified because if the subjects have failed to abstract the relevant rule in the training phase, then they will have to guess where the target will appear.

In contrast, if subjects respond on the basis of similarity to previously-seen sequences, then a different pattern should emerge. Table 5.2 gives the relevant predictions. Suppose the subject has seen the sequences shown in Table 5.1 in the training stage (minus the three sequences retained for the transfer phase) and is then tested with a sequence in which the stimulus appears in location C on trials 2 and 4. This sequence is quite similar to the two exemplars of the 'clockwise' and 'anticlockwise' rules that the subject has seen in training, but is less similar to the two exemplars of the 'same position' rule, in that it differs from each of the former by one location but differs from each of the latter by two. Thus we can predict that subjects should select either location B or C as their response, but not location A. Table 5.2 gives the predictions for the other test trials, together with the actual responses of the subjects.

It is clear that responding did not accord with the predictions of the rule-based account. For sequences that should conform to the 'same position' rule, location A was predicted less often than locations B and C; for sequences that should conform to the 'clockwise' rule, location B was predicted less often than locations A and C; and for sequences that should conform to the 'anticlockwise' rule, location C was predicted less often than locations A and B. As these results show, subjects chose the location that they had learned to be appropriate for previously-seen sequences, exactly as predicted by the instance-based theory.

Perruchet's elegant study should therefore serve to illustrate how these different types of behaviour may be discriminated. We have already discussed at length the idea that instance-based behaviour is determined by similarity to training items. For rule-based theories, on the other hand, similarity should play no role. This is because if old and new stimuli are treated alike, then new stimuli that are very dissimilar from training items should be responded to just as accurately as new stimuli similar to training items. The net effect is that inter-stimulus similarities will have no effect across members of the category; all that matters is how the relevant rule classifies the test item. We now turn to some studies which yield rather more positive evidence for rule learning.

Evidence for rule learning

Laboratory demonstrations of contrasts between rule and instance learning have been provided in a number of studies, and in this section we will consider a few of the best-known examples. We begin with some compelling evidence that has been reported by Lee Brooks and his colleagues (Allen and Brooks, 1991; Regehr and Brooks, 1993). The rationale of the experiments was as follows. Suppose that subjects learn to classify stimuli in a situation where a simple, perfectly predictive classification rule exists, and are then tested on transfer items that vary in similarity to the training stimuli. Observed behaviour to the transfer items can be of two contrasting types. On the one hand, 'bad' transfer stimuli similar to training items which the rule assigns to the opposite category may be classified as quickly and as accurately as 'good' items similar to training items the rule assigns to the same category. This would be consistent with classification being determined by the speeded application of a rule, where all that matters is whether the rule assigns the transfer item to one category or the other; whether the item is similar or not to a training instance, and whether that instance was in the same or a different category, should be immaterial. On the other hand, bad transfer items may be classified much less rapidly and accurately than good items, which would be consistent with categorisation on the basis of similarity to training instances; there would be no need to cite a rule as being part of the classification process.

Of course, we have already seen in studies like that of Homa *et al.* (Figure 3.7) that categorisation in some circumstances may be influenced by similarity to training items and hence not rule-based, but Allen and Brooks (1991) and Regehr and Brooks (1993) obtained evidence that both types of outcome can occur, depending on the type of stimuli used and the precise

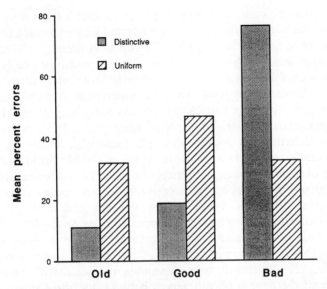

Figure 5.2. Percent classification errors for distinctive and uniform stimuli. Subjects learned to classify cartoon animals in a situation where a three-dimensional rule perfectly divided the two categories, and then were tested with training (old) items and 'good' and 'bad' transfer stimuli. Good and bad transfer items were highly similar to training items, but the good items were in the same category as the study items to which they were similar while the bad items were in the alternative category. For distinctive stimuli, bad items were classified much less accurately than good items, suggesting that classification was instance-based. For the uniform stimuli, good and bad items were classified with equal accuracy, suggesting that classification was rule-based. (After Regehr and Brooks, 1993.)

nature of the task. They trained subjects to classify animals into two categories ('builders' or 'diggers'). The animals varied in terms of five binary-valued dimensions: body shape, spots, leg length, neck length, and number of legs, but only three of the dimensions were relevant. The classification rule stated that category 1 was defined by the conjunction of long legs, angular body, and spots.

In one of Regehr and Brooks' experiments, subjects received 40 trials in the study phase on each of which one animal was presented and feedback was provided for the category decision. Then in the test phase old items were intermixed with new items that were either 'good' or 'bad'. Subjects were encouraged to respond quickly and accurately. Both good and bad items were highly similar to training stimuli, in that they differed on only one dimension, but the good items were in the same category as the study items to which they were similar while the bad items were in the opposite category.

There was one further manipulation in the experiment. For some subjects, the cartoon animals were highly distinctive in that the dimensions of variation of the stimuli were not interchangeable across stimuli. Thus half the animals had spots and half did not, but one animal's spots were different from another's. Similarly, half the animals had long legs and half short, but each set of long legs was slightly different. In contrast, for other subjects the stimuli were much more uniform, with the dimensions being interchangeable: spots for one animal were identical to those for another.

Figure 5.2 shows the key results from the experiment. Although not performing perfectly, subjects had clearly learned something about the category assignments of the training stimuli since at test the percentage of errors for old items was considerably less than 50%, which represents chance responding. Subjects made more errors on the old distinctive items than on the old uniform ones, presumably because the distinctive items are more individually memorable. For subjects shown distinctive stimuli, the original training items were classified best, the good test items somewhat worse, and the bad items were classified incorrectly on nearly 80% of occasions. This suggests that bad items were classified into the category of their nearest neighbour (which was, of course, in the alternative category). However, for subjects shown uniform stimuli, the pattern of results was quite different: there were no differences between the three types of test item, with the bad items in fact being classified slightly better than the good ones.

The implication of the results is that the specific dimensional structure of the stimuli controls whether they will be analysed into their component parts, which in turn determines whether rule- or instance-based classification will occur. While the decomposable uniform stimuli can readily be described and classified on the basis of a hypothesis, the distinctive stimuli lend themselves less well to description in terms of a rule. But perhaps subjects can be induced to classify the distinctive stimuli via a rule, given appropriate training conditions? In a further experiment, Regehr and Brooks trained and tested subjects in exactly the same way as in the previous experiment but told them at the outset what the classification rule was. Regehr and Brooks reasoned that telling the subjects the rule might increase the likelihood of rule-based classification, but Figure 5.3 shows that subjects continued to respond to the distinctive stimuli on the basis of similarity, since bad transfer items were again classified with more errors than good or old items. Overall, being told the rule allowed subjects to perform better than they otherwise would (errors are much less frequent than for the experiment shown in Figure 5.2), but it remained difficult for subjects to avoid computing similarity when classifying distinctive items. Uniform items again appeared to be readily classified according to the rule.

Taken together, these results suggest that under some circumstances rule-based classification is possible. It is not necessary to be given the objective rule in order to respond on the basis of it, nor is being given the rule a guarantee of rule-based behaviour.

Our criterion for rule-based behaviour is that classification latencies and

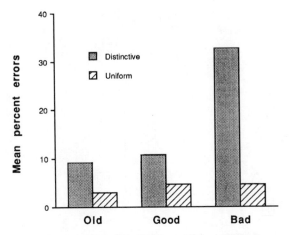

Figure 5.3. Percent classification errors for distinctive and uniform stimuli. Subjects again learned to classify cartoon animals in a situation where a three-dimensional rule perfectly divided the two categories, but in this case they were told the rule. Although the overall level of performance is much better than in Figure 5.2 (note the different scales on the ordinate), for distinctive stimuli, bad transfer items were again classified much less accurately than good items, while for uniform stimuli, good and bad items were classified with equal accuracy. Thus knowing the classification rule does not guarantee rule-based classification. (After Regehr and Brooks, 1993.)

errors should be no worse for novel test stimuli than for the actual training stimuli. The experiments discussed above reveal that there are situations in which these measures may be indistinguishable. However, while the observation of equal error rates for trained and novel stimuli is consistent with rule-learning, the converse does not necessarily hold: differences in error rates do not necessarily preclude rule-based performance. The criterion does not in fact require that performance to all of the test stimuli be identical: differences amongst the stimuli may still be consistent with rule-based classification.

To see how this may come about, consider an experiment by Nosofsky, Clark and Shin (1989). In this experiment, the stimuli were semicircles with an interior radial line: 16 stimuli were constructed from the combination of four sizes of semicircle (designated 1–4) with four angles of inclination of the radial line (again designated 1–4). Figure 5.4 shows the spatial layout of these quite confusable stimuli. The co-ordinates of the stimuli in the figure were derived from a multidimensional scaling analysis based on a confusion matrix generated by a group of subjects required to learn individual identification responses to the stimuli, in just the way discussed in Chapter 3 for Nosofsky's (1987) experiment.

In the classification part of the experiment, a different group of subjects received 300 trials in which they learned to classify 3 of the stimuli into category 1, with feedback, and 4 into category 2. Then in the test phase they

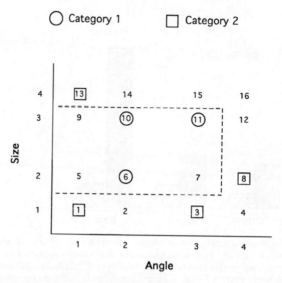

Figure 5.4. Category structure used by Nosofsky *et al.* (1989). The actual stimuli were semicircles varying in size with an interior radial line that varied in its angle, and the co-ordinates are from a multidimensional scaling analysis. Subjects were trained (with feedback) to classify stimuli 6, 10, and 11 into category 1 and stimuli 1, 3, 8, and 13 into category 2. The remaining stimuli are transfer items. The dotted line represents the boundary corresponding to a possible classification rule.

were required to make classification decisions (without feedback) to all 16 stimuli. Thus the dependent measures were the overall probabilities with which subjects placed each of the 16 stimuli into categories 1 and 2. In addition, in this experiment subjects were instructed to use the rule indicated by the dotted line in Figure 5.4 to classify the stimuli. That is to say, they were told at the outset that a stimulus is in category 2 if the value of size is 1, or the value of size is 4, or the value of angle is 4; otherwise the stimulus is in category 1. In the notation of set theory we can define category 2 as:

category 2: size=1 $\vee$ size=4 $\vee$ angle=4

Table 5.3 gives the probability across subjects that each stimulus was assigned in the transfer phase to category 1, and these probabilities confirm that something about the categorical structure had indeed been learned. With the exception of stimulus 15, items which the rule assigns to category 2 (1–4, 8, 12, and 13–16) were classified into category 1 with probability less than 0.5 (which means they were assigned to category 2 with probability greater than 0.5). In contrast, stimuli from category 1 were all assigned to that category with probability greater than 0.5.

Before considering the rule-based theory, it is worth first observing that the results appear to be quite contrary to the idea of instance-based classification. On this account, an item should be classified into the category

Table 5.3. *Results of Nosofsky et al.'s (1989) experiment*

	Stimulus	Probability	Mean
Category 1 training items	6	0.84	
	10	0.89	0.86
	11	0.86	
Category 2 training items	1	0.23	
	3	0.26	0.26
	8	0.15	
	13	0.38	
Test items (category 1)	5	0.81	
	7	0.84	0.81
	9	0.79	
Test items (category 2)	2	0.28	
	4	0.04	
	12	0.31	0.30
	14	0.44	
	15	0.54	
	16	0.16	

Test items are divided into those the rule assigns to category 1 and those assigned to category 2. The numbers refer to the probability of assigning a stimulus to category 1.

whose memorised exemplars it is more similar to, regardless of any rule that may discriminate the stimuli. But consider stimuli 4 and 7 in Figure 5.4. For both of these stimuli the nearest neighbours are stimuli 3 and 8, which are category 2 exemplars. While stimulus 4 is assigned, as expected, to category 2, stimulus 7 is placed in category 1 on over 80% of occasions. As Figure 5.4 shows, the rule puts these two stimuli into different categories, but it is very difficult to see how similarity to training items could achieve this. Likewise, it is hard to see how similarity could cause stimulus 16 to be assigned to category 2, when its nearest neighbour is stimulus 11, a category 1 exemplar. Again, the rule places these two stimuli in opposite categories.

If instance memorisation is not the basis of these decisions, then we must ask instead whether the classification probabilities are consistent with a rule-based account. On the assumption that subjects classify test stimuli according to the rule, we would expect that novel test stimuli would be classified just as accurately as the original study items. However, Table 5.3 shows that the observed classification probabilities differed widely across the stimuli, from which we might conclude that classification is not based on an underlying rule. In fact, the training stimuli from category 1 were classified with probability 0.86 while new items from that category were classified with a lower probability, 0.81. Similarly, new items from category 2 were more likely to be mistakenly assigned to category 1 (p=0.30) than were the original training items (p=0.26). This appears to discount the notion of rule-based classification, since our criterion requires no differences between old and new stimuli.

However, since the stimuli are perceptually highly similar, it is possible that, for example, stimulus 16 will be misrecognised as stimulus 11, in which case an incorrect classification decision will be made. Thus the probability of a category 2 response for stimulus 16 is a complex function. If it is correctly recognised as stimulus 16, then it will be assigned to category 2, since that is how the rule classifies it; if it is confused with any of the other category 2 stimuli, it will also be correctly classified; but if it is confused with any of the category 1 stimuli, it will be misclassified. However, since the scaling solution co-ordinates allows us to compute the relative similarities of all pairs of stimuli, it is possible to take into account the effects of these misrecognitions because we know the probability with which any given confusion will occur. Hence, we can predict the classification response for any given stimulus.

Figure 5.5 shows the observed classification responses versus the responses predicted from this more sophisticated rule model. The figure shows a very good match between observations and predictions, with 94% of the variance being accounted for. Figure 5.5 also shows the extremely poor fit to the data provided by the context model (41% of variance accounted for), indicating as expected that classifications are certainly not based on similarity to memorised training instances. Even with the possibility of selective attention to the size and angle dimensions allowing the inter-stimulus similarities to be altered, the context model is quite unable to explain the pattern of performance.

An even more compelling demonstration of the inadequacy of pure instance-storage comes from recognition memory data that Nosofsky *et al.* collected during the test phase of their experiment. When subjects were explicitly instructed to use a rule to classify the stimuli, no evidence emerged that the subjects could remember which test stimuli had been training stimuli. The probability of calling one of the training stimuli 'old' was exactly the same as the probability of mistakenly calling a novel test item 'old'. The implication of this result is that subjects had encoded nothing in the training stage except the rule: they had not memorised any of the instances. This persuasive finding shows that when appropriate conditions are established, subjects can indeed learn an abstract rule from exposure to instances.

Of course, just because Nosofsky *et al.*'s subjects were given rule-following instructions does not mean that this was necessary for rule-based behaviour to emerge. Indeed, we saw in Regehr and Brooks' study that rule-based

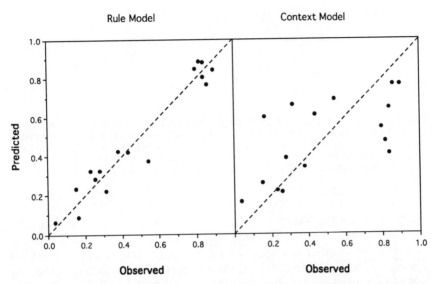

Figure 5.5. Observed versus predicted classification probabilities from Nosofsky *et al.*'s (1989) experiment. Each point refers to one of the 16 stimuli from Figure 5.4, and the axes show the probability of a category 1 response. The left-hand panel gives the predictions of the rule-based model, and the right-hand panel gives the predictions of the context model. Subjects' classification responses are much better accounted for by the rule-based model. (After Nosofsky *et al.*, 1989.)

behaviour can emerge without specific instructions regarding the rule, but whether the same would be true for Nosofsky *et al.*'s procedure is unknown. In any case, the stimuli Nosofsky *et al.* used probably make rule-following difficult in that they do not readily lend themselves to verbal descriptions.

A final piece of evidence for abstraction comes from so-called 'function learning' experiments demonstrating that subjects can make accurate extrapolations concerning novel stimuli. Suppose a subject learns to make a range of unidimensional responses, $R_1...R_n$, to a range of unidimensional stimuli, $S_1...S_n$. The stimuli might be lines of varying length and the responses button-presses of varying durations. If the subject is then tested on an extrapolation trial with stimulus S_{n+1}, an instance theory will predict that he or she should make the response appropriate for the training stimulus that is most similar to S_{n+1}, namely R_n. Remember that according to the context model, when a novel stimulus is presented it is assigned a response that has been associated with a similar previously-seen stimulus. On this account, it is hard to imagine how entirely novel responses could be generated.

In fact, as DeLosh (1993) and Koh and Meyer (1991) have shown, subjects are quite good at making novel responses (i.e., R_{n+1}) to novel test stim-

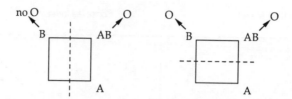

Figure 5.6. Rule-based account of contingency effects. Left: a contingent relationship between cue A and outcome O resulting from AB→O and B→no O trials. The dotted line represents the rule that all stimuli possessing cue A are associated with outcome O. Right: a noncontingent relationship between cue A and outcome O resulting from AB→O and B→O trials. The dotted line represents the rule that all stimuli possessing cue B are associated with outcome O.

uli, but this is a phenomenon that seems to lie outside the scope of current instance theories. The ability to extrapolate seems to depend on the formation of some abstract rule specifying the general relationship between the stimuli and responses, such as 'as the stimulus increases in length the correct response increases in duration'. Thus in DeLosh's experiment, subjects were able to press a button for a duration that was approximately appropriate for a novel line length. We would either have to conclude that it is inadequate to claim that associative learning is in general based on the memorisation of instances (with responding being governed by similarity to those instances), or else cite some process whereby novel responses can be generated. Contradicting instance theories, the evidence strongly suggests that subjects can learn the abstract functions that relate stimuli to responses.

The overall conclusion from this section, then, is that people can learn a rule or hypothesis that maps stimuli onto outcomes and that this way of acquiring associative knowledge can be dissociated from other ways such as instance-memorisation. In the next section we will look at some evidence for abstract representations that go beyond the specific training stimuli, but first it is important to note that we now have an alternative account of the effect of contingency on associative learning. The rule-based explanation is simply that the subject learns a different rule when the relationship is contingent compared to when it is noncontingent. The left panel of Figure 5.6 illustrates a contingent situation in which outcome O accompanies AB trials but not B trials. Here, the rule represented by the vertical dotted line divides the space into two regions, consisting of all stimuli such as AB that predict the outcome and all stimuli such as B which do not. Clearly, on this rule the outcome will be expected when cue A occurs on its own. In contrast, the right-hand panel of the figure shows a rule for dividing the space when both AB and B are paired with the outcome. Now, the outcome will not be expected when cue A alone occurs.

In addition to being able to explain the basic effect of contingency, these rules have a rather natural psychological interpretation in terms of necessary and sufficient conditions. In the contingent case, the rule essentially says that cue A is both necessary and sufficient for the production of the outcome, while in the noncontingent case it is cue B that has this status.

Narrow and broad abstract representations

Because a rule applies equally to novel and to old stimuli, learning a rule requires the formation of an abstraction. In contrast to the simple encoding of instances, where nothing need be learned beyond the training stimuli themselves, rule learning involves the acquisition of knowledge which transcends the training stimuli. To return to our example concerning rectangles varying in height and width, the rule (that rectangles higher than they are wide are in one category and rectangles wider than they are high are in another) is an induction that applies to far more than just the training items from which the rule was inferred. Such an abstraction process has the great benefit of representing the category structure in a very simple way, in contrast to the very memory-intensive requirements of an instance-storage process. Although it may be a good deal harder, in terms of active information processing, to induce a rule than to memorise a set of instances, the end product is usually a very compact and efficient representation.

Thus far, the evidence for rule learning that we have considered requires only a fairly narrow degree of abstraction. What I mean by this is that the rules we have discussed can apply to novel stimuli characterised by the same feature dimensions as the training items. Take Nosofsky *et al.*'s (1989) data for example. Here, we have evidence that as a result of exposure to one set of semicircles with internal radial lines, subjects can learn a rule which applies to other, novel, semicircles with internal radial lines. But we have no reason to believe that Nosofsky *et al.*'s subjects would have been able to apply their rules to entirely different stimuli, such as triangles or squares. To do this, one needs to form what I shall call a 'broad' rather than a 'narrow' abstraction. In this section we will consider various lines of evidence that subjects can learn broad as well as narrow abstractions in associative learning tasks.

We begin by considering some evidence from the learning of artificial grammars. Our extensive considerations in previous chapters led us to the conclusion that, in the main, what a subject does in the study phase of an artificial grammar learning experiment is to memorise the training strings. Recall that the basic procedure is to expose the subject to a number of letter strings generated from a grammar and then to ask him or her to try to discriminate between new grammatical and nongrammatical strings. The ability to do this at better-than-chance levels is explained by the sorts of instance memorisation processes inherent in connectionist networks.

There is, however, some evidence that subjects may do more than simply encode unanalysed training strings; in fact, they may learn very broad rules. Consider a study by Brooks and Vokey (1991) in which subjects were trained and tested on strings based on different sets of letters. After being trained on one set of strings generated from the grammar, subjects were then tested on strings generated from the same grammar but in which all the letters had been changed (e.g., M→Q, V→Z, etc.). The importance of this procedure is that above-chance transfer would seem to rule out the sort of instance-based process that we have been considering, at least as a complete explanation of the findings. If the subjects have simply memorised study strings like MXRVXT, and respond to test strings on the basis of their similarity to the encoded strings, it is hard to see how an item like QJLZJF can be classified as grammatical, because its similarity to the study item is so low. Instead, the ability to respond 'grammatical' to this item would seem to require some form of broad, abstract, rule-based knowledge, for instance knowledge that 'if an item has the same letter appearing in the 2nd and 5th positions, it is grammatical'. This abstraction is broad in the sense that it applies to stimuli characterised by entirely different features (letters) from the study items.

What were the results for this group of subjects? Brooks and Vokey obtained above-chance transfer performance, with 55.5% of test items being correctly classified, a value that is significantly greater than chance but also below that achieved by subjects tested on strings from the same letter set as the study items. The latter result argues against the view that all knowledge is abstract, since in that case the change of letter sets should make no difference. More importantly, the fact that subjects could still perform above chance strongly challenges the view that similarity to memorised exemplars is a sufficient basis for responding, and instead argues that at least some abstract knowledge had been acquired.

However, to judge whether the value of 55.5% represents reliable transfer or not, we need to compare it not with the chance value of 50.0%, as Brooks and Vokey did, but rather with the obtained percent correct in a group of subjects performing the grammaticality judgment task without having witnessed the study phase. While subjects in such a task have no study information to guide their decisions, it is by no means obvious that they will perform at the chance level of 50.0% correct because within the test itself some degree of learning may be possible. Brooks and Vokey did not run such a control group, but fortunately an experiment by Altmann, Dienes, and Goode (1995) provides just such a contrast, and goes further by showing that subjects can transfer knowledge not just between different letter sets, but also between different modalities.

Altmann *et al.* used the same basic design as Brooks and Vokey, except that instead of transferring subjects from strings instantiated against different letter sets, subjects transferred from letter strings to strings of tones or vice versa. One group of subjects was trained on a standard set of grammat-

Table 5.4. *Results of Altmann et al.'s (1995) experiment*

	Learning set		
	None	Letters	Tones
Test set			
Letters	49.7	59.2	53.7
Tones	48.5	55.8	57.3

Figures are percent correct classifications.

ical letter strings. For another group, the strings were formally identical but consisted of sequences of tones. Thus each letter string had a counterpart tone sequence in which, for instance, the letter M was translated into a tone at the frequency of middle C. In the test phase, subjects either made grammaticality judgements for the same stimuli they saw in the study phase (trained letters/tested letters or trained tones/tested tones) or for a different type of stimulus (trained letters/tested tones or trained tones/tested letters). Finally, some further subjects were tested without having been exposed to any training items. For these subjects, performance on the test should be close to chance unless some degree of learning is possible in the test itself.

The results are shown in Table 5.4. First, note that subjects not given any study trials performed very close to chance (50% correct) in the grammaticality test, confirming that they were unable to learn the features that discriminate grammatical from nongrammatical items during the test itself. Next, note that subjects trained and tested on letter strings or trained and tested on tone sequences performed well above chance. We have seen the result from the former group before, but the result from the latter group shows that the ability to make well-formedness decisions on the basis of exposure to a sample of items from a grammar is not exclusive to letter strings. Finally, the most interesting result is that subjects trained on letters and tested on tones or trained on tones and tested on letters were also able to perform above chance. While the mean level of accuracy (54.7% correct) is lower than that of the groups trained and tested in the same format (58.2% correct), it is nevertheless significantly better than that achieved by the control subjects.

The clear implication of this study is that while exposure to a set of grammatical items may lead to instance memorisation, it also engenders a degree of broad abstraction such that stimuli from the same grammar, but instantiated in a totally different format, can be distinguished from nongrammatical items. Although the benefit seen when subjects are trained and tested with items in the same format suggests that instance memory continues to play a role, it is plainly not sufficient to assume that memorisation is the

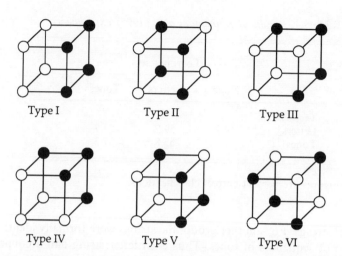

Figure 5.7. Stimulus structures used by Shepard *et al.* (1961). If there are eight stimuli varying on three dimensions, then 70 possible partitions can be formed in which four stimuli belong to each category. However, these 70 partitions can be reflected or rotated into the six structures shown in this figure. In a type I classification, only one dimension is relevant, while in a type VI classification all dimensions are relevant. Filled vertices correspond to members of one category, unfilled vertices to those of the other category.

only process taking place at the time of learning. Of course, Altmann *et al.*'s results do not tell us what sorts of rules subjects might have been learning, and subtle experiments will be required to answer that question. It is by no means necessary that the rules be sophisticated: maybe they consist of very simple hypotheses such as 'the first two items of the sequence cannot be the same'. But bear in mind that the amount of time given to subjects to learn about the structure of the stimuli was very brief: extended training would, in all probability, allow subjects to learn much more about the rules of the grammar.

Another line of evidence for broad abstract representations comes from a classic experiment by Shepard, Hovland and Jenkins (1961). Shepard and his colleagues were interested in the relative ease or difficulty of learning various categorisation problems which are identical in all respects except for the underlying category structures. Suppose that we have eight stimuli varying on three dimensions: they may be circles or squares that are large or small and green or red. If we assign four of these stimuli to one category and four to the other, then there are a large number of different partitions that may be constructed, but it turns out that these are all reducible to six basic structural types, and these are shown in Figure 5.7. All of the numerous (in fact, 70) ways of partitioning the stimuli can be reflected or rotated into one of these six types. In this figure, filled vertices correspond to members of one category and unfilled vertices to members of the other.

Shepard *et al.* trained subjects to classify stimuli into the appropriate categories and found a very clear ordering of the relative difficulty of learning each type. Type I was easiest, followed by type II, then types III, IV, and V (which were about equal in difficulty), and finally type VI. How can we explain this ordering? A glance at the category structures reveals that the ordering is consistent with rule complexity. A type I rule only requires reference to one dimension, which means that an adequate rule might be:

C1: red

which states that a stimulus is in category 1 if it is red (otherwise it is in category 2). Clearly, only the colour dimension needs to be analysed in order to apply this rule. A type II rule, on the other hand, requires reference to two dimensions, as in:

C1: (triangle and red) or (square and green)

which states that a stimulus is in category 1 if it is a red triangle or a green square (otherwise it is in category 2). Both colour and form need to be known in order to apply this rule. The hardest rule, type VI, requires reference to all three dimensions. Thus it is plausible to speculate that the observed ordering may come about because the subjects are attempting to infer a rule for category membership, and the ease of inducing this rule is different for the different structures.

Whatever the merits of such an explanation (see Nosofsky *et al.* 1994, for some alternative accounts), additional data that Shepard *et al.* obtained seem to provide clearer evidence that some degree of broad abstraction was occurring in the study phase. In addition to comparing the ease of learning of these problems, Shepard *et al.* examined transfer of rule learning. For each type of classification, subjects learned five classifications in succession, all with the same underlying rule but differing in the actual stimuli used: a new set of stimuli was presented for each of the five classifications. Thus in one problem the stimuli might have varied on the three dimensions described above (size, colour, shape), in another the stimuli might have consisted of three objects consisting of a musical instrument (violin/trumpet), light (candle/bulb), and mechanical item (nut/screw). Although the specific stimuli used in different problems were quite different, the underlying structure was maintained. For each problem, the eight training stimuli were presented in a random order for classification, and feedback was provided to tell the subject what the correct response for each stimulus was. Training continued until 32 consecutive correct responses were made.

Shepard *et al.* obtained very large transfer effects. In type VI problems, over 60 errors were made on average before the learning criterion was reached on the first problem, but only 20 errors were made on the fifth problem. Since the problems used different stimuli, transfer could not have involved generalisation to memorised instances, nor could abstraction of a narrow kind have been responsible: although the dimension of colour might have been relevant in one problem, the next problem would have used quite different stimulus dimensions. Instead, subjects appear to have been able to learn much broader abstractions. For instance, to show some benefit on a type VI problem as a result of prior training on another problem of that sort (but instantiated with quite different stimuli), the subject would have to carry over a rule such as 'If a given stimulus is in category A, then any stimulus differing from it by one feature is in category B, any stimulus differing by two features is in category A, and any stimulus differing by three features is in category B'. Of course, it is unlikely that Shepard *et al.*'s subjects had learned this rule in its entirety given only a few problems of each type; instead, they probably learned only some part of it, and different subjects may have learned different parts.

Nevertheless, it is possible that with extended training subjects might be able to learn and carry over the complete rule. What behaviour would then emerge? In the case of a type VI problem, if the subject has genuinely induced the entire rule, then when presented with a new problem instantiated with novel stimuli, it should only be necessary for him or her to observe one of the stimulus-category assignments to be able to infer the rest. Is there any evidence of such one-trial learning? Numerous studies with humans and animals have looked at the formation of 'learning sets', and results exactly in accordance with the above expectation have been obtained. For instance, suppose an organism is shown a pair of objects, A and B, and is rewarded for choosing A but not B. After a few trials, the reward contingency is mastered. Then, a pair of new objects C and D is introduced, with choice of the former being rewarded. This procedure is repeated many times with new pairs of objects E and F, G and H, and so on. The result of such an experiment has been well-known since Harlow's (1949) pioneering studies with rhesus monkeys. Eventually, each new problem is learned in a single trial. Of course, on trial 1 the subject must guess, but having seen whether the chosen stimulus is rewarded or not, performance from trial 2 will be close to perfect. Thus with sufficient exposure, subjects appear to be able to abstract a rule of the form 'if one stimulus is rewarded then the other is not'.

An even more compelling example comes from experiments demonstrating the formation of so-called 'equivalence sets'. Suppose subjects are taught that some stimuli (the A set) are paired with a reward such as food while other stimuli (the B set) are nonreinforced. The subjects then see a reversal in which the B stimuli are reinforced and the A ones not, then a reversal back to the original contingency, and so on for many reversals. At the end of one reversal phase, the animal has learned, say, that the A stimuli are rewarded and the B ones not. Then at the beginning of the next reversal, to the subject's surprise one of the B stimuli is reinforced. Will the subject be able to infer that the reward contingency has now changed across the whole set? That is, will it now expect reward following the remaining **B** stimuli and not following the **A** stimuli?

The answer appears to be yes. In appropriate circumstances, a whole set of stimuli can come to be treated as equivalent, such that when something new is learned about one of the stimuli, that knowledge automatically transfers to the remaining ones. This can occur not only in humans but in other animals too. As an example, in an experiment with pigeons, Vaughan (1988) took a collection of 40 photographs of outdoor scenes and divided them at random into two sets such that there was no obvious feature or set of features that distinguished the sets. A slide was presented on each trial, with half of the slides being followed by food and half being nonreinforced. Since the pigeons were hungry, it is reasonable to assume that they would try to work out what it was that characterised the reinforced slides. Over many trials, pigeons learned to respond to members of one set but not to members of the other. When stable responding had been established, a series of reversals occurred, each lasting long enough for the pigeons to learn the new reinforcement arrangement. In each reversal, all of the stimuli that had previously been paired with food were now presented without food and vice versa. After many such reversals, the pigeons only had to see one or two slides at the beginning of a reversal to be able to work out the reward contingencies for all of the remaining stimuli. Thus members of the two sets were treated as equivalent, which fulfils our requirement for rule-based responding. In situations of this sort, groups of arbitrary stimuli are mentally represented as sets, such that a property that comes to be associated with one member of the set is immediately inherited by all of the other members.

In humans it is possible to develop the notion of equivalence rather more formally. Mathematically, the relationship R between objects is an equivalence one if three simple conditions are met, called reflexivity, symmetry, and transitivity. First, it must be the case that each object bears relationship R to itself (reflexivity, denoted aRa, bRb). Secondly, the relationship must be symmetric, such that if a holds relationship R to b, then b holds the same relationship to A (if aRb, then bRa). Finally, the relationship must be transitive, such that if a and b are in relationship R, and b and c are in relationship R, then a and c are in relationship R (if aRb and bRc, then aRc). Applying these criteria to the mathematical relationship of equality (=), we can see that each of the following is true, and so equality implies equivalence:

Reflexivity: $a=a$
Symmetry: if $a=b$ then $b=a$
Transitivity: if $a=b$ and $b=c$ then $a=c$.

Numerous experiments have now shown that humans can learn to treat sets of stimuli in such a way that the criteria of equivalence are held. In their

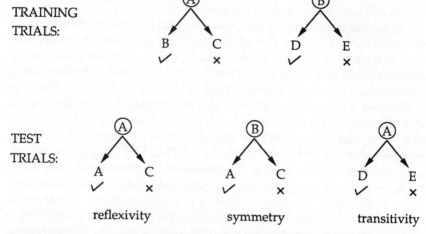

Figure 5.8. Training procedures for the formation of equivalence sets. In the training phase, stimulus B is the correct choice in the presence of conditional cue A (on different trials, B would appear on the right and C on the left), and stimulus D is the correct choice in the presence of B. As a result of this training, stimuli A, B, and D come to be treated as equivalent. This is shown by the fact that in the test stage, subjects choose stimulus A in the presence of conditional cue A (reflexivity), choose A in the presence of B (symmetry), and choose D in the presence of A (transitivity).

pioneering experiments on equivalence, Sidman and Tailby (1982) used a conditional discrimination procedure in which one stimulus signalled which of two choice responses was rewarded. People were trained to choose stimulus B rather than stimulus C in the presence of conditional stimulus A and to choose D rather than E in the presence of B, as shown in Figure 5.8. Subsequently, tests established that stimuli A, B, and D were treated as equivalent. Subjects chose A in the presence of A (reflexivity), chose A in the presence of B (symmetry), and chose D in the presence of A (transitivity).

Such results establish that people can learn to treat unrelated stimuli as equivalent in such a way that one stimulus can stand for another. Plainly, learning and generalising between memorised training items would be completely inadequate for this process, because such a process would be unable to explain why a new response learned to one stimulus generalises to its equivalent.

Verbal reports

There is one final source of evidence from subjects' behaviour that seems consistent with the notion that rules can readily be abstracted from training stimuli. Probably the most persuasive evidence for rule-induction as a mechanism for the learning of some concepts is the evidence from subjects' verbal reports. Subjects are often able to say what rule or hypothesis they are currently considering and to justify their behaviour with reference to a rule. Introspectively, we are often aware of following a rule, as for example when we follow some directions to get to a particular location.

I mentioned earlier that the notion of a rule has been a topic of immense controversy in psychology and philosophy. The controversy has revolved around how one should answer two sorts of questions. Apart from deciding what sort of behaviour we would expect to see from rule-following organisms, we also need to know what additional requirements there are for us to be sure that a rule is governing behaviour. Without going into any of the complexities, the most widespread view is that the person needs to be able to justify their behaviour in terms of the putative rule. Imagine I have a black box which is capable of learning to classify objects presented to it, and suppose that all the evidence suggests that the system is making its decisions according to an induced rule. According to philosophers such as Wittgenstein (1958) and Kripke (1982), we cannot attribute rule-following behaviour to our black box unless it is able to give an appropriate normative justification for each of its decisions. Such a justification would take the form of saying something like 'I classified the object as an X because it possessed feature F and I was following rule R'.

Whether or not one is convinced by this requirement, it is clear that reasons and justifications lend extra weight to claims about rule-learning. Experimental evidence confirms that verbal reports provide a useful index of rule-based behaviour. For example, in addition to examining transfer to novel stimulus sets, Shepard *et al.* (1961) required their subjects to verbally report, for each of the five problems of each type, how they were making their classifications. Most subjects reported that they were using rules, and in addition, in most of these cases the rules were correct (in the sense that they correctly assigned stimuli to categories). But many of the stated rules were overly complex. For instance, the simplest rule for the type VI problem shown in Figure 5.7 (assuming the stimuli are circles or squares that are large or small and red or green) would say that a stimulus is in category 1 if it is a circle and is large and is red, or if it is just one of these; otherwise it is in category 2. A much more complicated rule for this problem would simply enumerate all of the stimuli: a stimulus is in category 1 if it is a large red circle, a large green square, a small green circle, or a small red square, otherwise it is in category 2. Clearly, this is a less compact rule than the first one.

What would be the consequence of using an overly-complex rule? Plainly, since a person's ability to manipulate rules is going to be affected by the limits of short-term memory, it is likely that errors will be far more common in subjects using complex rules than in those using simpler ones. To test this, Shepard, Hovland, and Jenkins rated each of the rules that the subjects verbalised in terms of its complexity. Unsurprisingly, they found that the rules were often overly complex for the first problem of each type, but that they became more economical as further problems of the same type were tested. For instance, a subject might solve the first type VI problem by stating a rule which simply enumerated the stimuli, but on the final problem of this type might report the compact rule given above. Most importantly, Shepard *et al.* found that the number of classification errors subjects made was highly correlated with rule complexity. If the subjects were basing their classifications on these reported rules, this is exactly what we would expect: trying to classify stimuli on the basis of a rule that is too complex to keep in mind is likely to lead to errors.

Many people would probably regard it as controversial to claim that rule-based behaviour can only be inferred if subjects are able to provide adequate verbal reports about their behaviour, since such a condition would seem to preclude the possibility of rule-based behaviour in animals and non-verbal humans. Nevertheless, it is clear that reasons and justifications lend extra weight to claims about rule-learning. The fact that Shepard *et al.*'s (1961) subjects gave verbal justifications in close accordance with their behaviour is therefore highly significant.

Relational concepts

In most of the cases we have considered thus far, rule induction requires the existence of relational concepts. To see what I mean by this, let us return once again to the example of rectangles varying in height and width. In order to decide whether a sample stimulus falls into the target category or not, we need to be able to judge whether its height is greater or less than its width. But there is no simple perceptual feature, or set of features, whose presence or absence will tell us the answer to this question. Instead, we need to make a comparison between two features (height and width). This may seem trivially obvious, but it is important to note that the ability to make this judgement presupposes the possession of relational concepts such as 'greater than'.

We tend to take the possession of such concepts for granted, but their importance can be illustrated by considering a task that is trivially easy for humans to learn but which would be beyond the capabilities of an organism lacking the simplest of relational concepts. Herrnstein *et al.* (1989) conducted a well-known categorisation experiment with pigeons in which a slide was presented on each trial, with half of the slides being followed by food and half being nonreinforced. The essential property was not any single perceptual feature or set of features, but rather the presence or absence of the abstract relation of 'insideness'. Specifically, each slide consisted of a dot that was either inside or outside a closed figure, and food accompanied all of the 'inside' stimuli. Because of the way the stimuli were constructed, there were no particular perceptual features or sets of particular features that correlated with food; for example, the 'outside' stimuli did not occupy a larger region of the slide than the 'inside' stimuli. Rather, the discrimination could only be learned by judging whether or not the dot was inside the figure.

The key question is whether the pigeons were able to master this classification problem. It is obvious that it would be trivial for humans. Children as young as four are known to have relational concepts such as insideness, sameness, and so on (e.g., Smith, 1989). In fact, Herrnstein *et al.*'s pigeons only showed very weak evidence of being able to learn the classification, suggesting that 'insideness' is beyond their conceptual capability. What is it that allows humans but not pigeons to master this sort of problem? Clearly, the simplest answer is that possession of genuine relational concepts like 'insideness' is the critical factor. It must be possible, given a stimulus array, to extract from it certain abstract features such as that one part is inside another, one element is different from another, and so on. The associative learning capabilities of humans are plainly boosted by the ability to interpret stimuli in terms of such concepts.

We are naturally led to wonder where such concepts come from. This book is not the place to consider this topic in any detail, but it is worth noting that possession of such concepts represents a considerable challenge to the empiricist programme that has been generally advocated in this book, since it is hard to see how exposure to positive and negative examples of a concept like 'insideness' could ever be sufficient for that concept to be acquired, no matter how powerful the learning mechanism. To learn such a concept seems to require the ability to represent the stimulus on a dimension (inside-outside) that presupposes the very concept that we are trying to explain.

Mechanisms of rule-learning

Having established that associative learning may in some circumstances be based on the induction of rules, and that these rules may include relational concepts in their specification, our final topic is to briefly consider the sort of learning mechanism that may underlie rule-learning. Since the predictions of rule- and instance-based accounts can be so divergent, it may seem that quite different mechanisms will be needed. Clearly, our explanation of instance-based behaviour will be in terms of the sort of connectionist mechanisms discussed in the last chapter. But do we really need something quite different to explain rule-based behaviour, or is it possible that a single mechanism could underlie all these cases? Is it possible that apparent rule- and instance-based behaviour may in fact both emerge from a connectionist system?

This is an important question but regrettably one which has received rather little attention, so in the present section I will simply try to sketch some of the major problems that need to be addressed. Let us begin by assuming that a connectionist system, such as Kruschke's ALCOVE model, can provide a reasonably good account of instance-based behaviour. How would such a model need to be modified in order to explain the results discussed here? First, consider our example of rectangles varying in height and width. If exposed to a number of examples, some of which are wider than they are high and which belong to category A, others of which are higher than they are wide and which belong to category B, then we know that the system will be able to learn the classification, but will do so by essentially memorising the training items and responding to novel items as a function of their similarity to the memorised stimuli. That is to say, responding will be graded and hence not rule-based.

But a modest intervention could allow such a connectionist system to behave appropriately. All that is required is that there be a unit in the network (either a hidden or an output unit) which integrates information about the two input dimensions, and which applies a sharp threshold at the appropriate place (see Figure 5.9). Thus suppose we have a unit which receives input about height and about width and which generates a positive output whenever height is greater than width and a negative output whenever width is greater than height. This can be achieved, as shown in Figure 5.9, by arranging for the weight from the 'height' input unit to be +1 and the weight from the 'width' input unit to be –1. Then all that is required is that the unit have an activation function of the sort shown in the figure, where a positive output (say +1) is generated whenever the input is greater than zero and a negative output (–1) when it is less than zero. The net result is that the system produces an output of –1 for all members of category A, regardless of whether they are new or old and regardless of how similar to training stimuli they are, and an output of +1 for all members of category B.

Such a network is clearly behaving as if it is following a rule. What about situations in which the rule is more complex? Here, as in the case of Nosofsky *et al.*'s (1989) design, the network would need to have several appropriately-tuned units in order to deal correctly with all input stimuli, but in principle rule following behaviour could again emerge. Thus the basic phenomenon of rule-based behaviour – equivalent responding to new and old stimuli – is not per se at variance with the processing capabilities of connectionist systems. Where such systems are limited, though, is in terms of actually learning to behave in a rule-based manner. It is one thing for the experimenter to hand-design a network to show a particular sort of behaviour, but it is quite another for the system to manifest that behaviour simply as a result of learning. As we know, if left to learn for itself, a network is rather unlikely to develop a weight structure capable of reproducing the rule-based behaviour of, say, Nosofsky *et al.*'s subjects. Moreover, connectionist learning systems depend on the provision of feedback associated

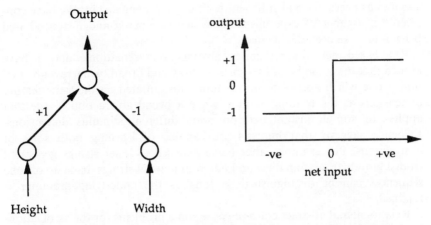

Figure 5.9. A simple network capable of demonstrating rule-based responding. Suppose the training stimuli are rectangles, and category A consists of rectangles wider than they are high and category B consists of ones higher than they are wide. One input unit receives activation proportional to the height of the rectangle, the other receives activation proportional to the width. These activations are then multiplied by weights of +1 and −1, respectively. The net input to the output unit is therefore negative for category A stimuli and positive for category B stimuli. If the output unit has an activation function of the form shown in the graph on the right, it will produce the same output (−1) for all A stimuli, and likewise (+1) for all B stimuli. This will occur for novel as well as training stimuli, and so the system is effectively following a rule.

with positive and negative examples of a category. Humans, by contrast, can learn a rule merely via linguistic instruction, without seeing any examples at all.

A further problem is that even when rule-based behaviour emerges from a connectionist system, it is likely to involve only narrow abstraction to stimuli instantiated with the same feature dimensions. Of course, the real problem is broad abstraction, because that requires abstract predicates. Where would the network get these? At present it is hard to see how such representations could ever be learned, but because of their complexity it is dangerous to assert that connectionist networks definitely will or will not be able to learn certain things. What we can be sure of is that a great deal of further work is required to see whether connectionist networks have anything useful to say about how humans acquire abstract concepts.

Summary and overview

In this chapter I have described a number of experimental results that seem to require an explanation in terms of a rule-induction process. Such a process involves the abstraction of a classification rule that treats new and old stimuli alike, and determines the category an item belongs to simply on the basis of whether the stimulus possesses the attributes which the rule specifies as necessary and sufficient for category membership. We have considered a number of experiments in which rule-based and instance-based behaviour have been dissociated.

Rule-based behaviour requires abstract representations, and I have argued that these can be of two sorts, narrow and broad. A narrow abstraction is one which applies to novel stimuli instantiated on the same feature dimensions as the training stimuli, while a broad abstraction is one that applies to stimuli instantiated on quite different stimulus dimensions. Evidence suggests that humans are capable of learning both sorts of abstractions. For example, after being exposed to letter strings generated from a grammar, grammatical and nongrammatical strings made up of tone sequences can be distinguished so long as the underlying grammar is retained.

Relational and abstract concepts pose some problems for the associationist approach that has been advocated in this book. While there can be little doubt that such concepts play a role in associative learning, we will have to await further research before judging whether connectionist models can illuminate the mechanisms of rule learning.

In the context of the overall view of associative learning I have presented in this book, the place of rule induction is a slightly uncomfortable one. I started out by considering three different but interrelated questions one could ask about learning. The first concerns the extent to which learning is rational in the sense of conforming to a normative theory. Given an appropriate normative theory, developed in Chapter 2, it emerges that judgments of association are often highly accurate, and deviate from the normative theory no more than one would expect of any system that has evolved fundamentally to exploit the causal and structural regularities of the world. Our second question concerned the informational substrate of learning. In Chapter 3, we saw that an enormous amount of data can be understood in terms of the memorisation of instances, with responding to novel items being a function of their similarity to the ensemble of stored instances. Finally, in Chapter 4 I described how contemporary connectionist models tackle the third question: at the mechanistic level, how is associative knowledge acquired? Connectionist network models provide machinery for the development of inhibitory and excitatory connections between elements and for the transmission of activation from one representation to another.

Rule learning does not fit very easily within this scheme. On the one hand, some of the evidence reviewed in the present chapter seems to imply that rule induction relies on a system quite unconnected to the instance memorisation machinery. On the other hand, it is possible that rule learning will eventually yield to an explanation in terms of connectionist processes. We can be sure that investigation of the ways in which these different processes for acquiring associative knowledge are related to one another will form a major aspect of future research.

References

Allan, L. G. (1980). A note on measurement of contingency between two binary variables in judgment tasks. *Bulletin of the Psychonomic Society*, **15**, 147–9.

Allan, L. G. & Jenkins, H. M. (1983). The effect of representations of binary variables on judgment of influence. *Learning and Motivation*, **14**, 381–405.

Allen, S. W. & Brooks, L. R. (1991). Specializing the operation of an explicit rule. *Journal of Experimental Psychology: General*, **120**, 3–19.

Alloy, L. B. & Abramson, L. Y. (1979). Judgment of contingency in depressed and nondepressed students: sadder but wiser? *Journal of Experimental Psychology: General*, **108**, 441–85.

Altmann, G. T. M., Dienes, Z. & Goode, A. (1995). On the modality-independence of implicitly learned grammatical knowledge. *Journal of Experimental Psychology: Learning, Memory, and Cognition*. (In press.)

Anderson, J. A. (1968). A memory storage model utilizing spatial correlation functions. *Kybernetik*, **5**, 113–19.

Anderson, J. R. (1990). *The adaptive character of thought*. Hillsdale, NJ: Lawrence Erlbaum Associates.

Anderson, J. R. & Schooler, L. J. (1991). Reflections of the environment in memory. *Psychological Science*, **2**, 396–408.

Ashby, F. G. & Gott, R. E. (1988). Decision rules in the perception and categorization of multidimensional stimuli. *Journal of Experimental Psychology: Learning, Memory, and Cognition*, **14**, 33–53.

Attneave, F. (1957). Transfer of experience with a class-schema to identification-learning of patterns and shapes. *Journal of Experimental Psychology*, **54**, 81–8.

Baker, A. G., Mercier, P., Vallee-Tourangeau, F., Frank, R. & Pan, M. (1993). Selective associations and causality judgments: presence of a strong causal factor may reduce judgments of a weaker one. *Journal of Experimental Psychology: Learning, Memory, and Cognition*, **19**, 414–32.

Barsalou, L. W. (1990). On the indistinguishability of exemplar memory and abstraction in category representation. In T. K. Srull & R. S. Wyer (Eds.), *Advances in social cognition* (Vol.3, pp. 61–88). Hillsdale, NJ: Lawrence Erlbaum Associates.

Beach, L. R. & Scopp, T. S. (1966). Inferences about correlations. *Psychonomic Science*, **6**, 253–4.

Bitterman, M. E., Tyler, D. W. & Elam, C. B. (1955). Simultaneous and successive discrimination under identical stimulating conditions. *American Journal of Psychology*, **68**, 237–48.

Broadbent, D. E. (1985). A question of levels: comment on McClelland and Rumelhart. *Journal of Experimental Psychology: General*, **114**, 189–92.

Brooks, L. R., Norman, G. R. & Allen, S. W. (1991). The role of specific similarity in a medical diagnostic task. *Journal of Experimental Psychology: General*, **120**, 278–87.

Brooks, L. R. & Vokey, J. R. (1991). Abstract analogies and abstracted grammars: comments on Reber (1989) and Mathews *et al.* (1989). *Journal of Experimental Psychology: General*, **120**, 316–23.

Bruner, J. S., Goodnow, J. J. & Austin, G. A. (1956). *A study of thinking*. New York: Wiley.

Chandler, C. C. (1991). How memory for an event is influenced by related events: interference in modified recognition tests. *Journal of Experimental Psychology: Learning, Memory, and Cognition,* **17,** 115–25.

Chandler, C. C. (1993). Accessing related events increases retroactive interference in a matching recognition test. *Journal of Experimental Psychology: Learning, Memory, and Cognition,* **19,** 967–74.

Chapman, G. B. (1991). Trial order affects cue interaction in contingency judgment. *Journal of Experimental Psychology: Learning, Memory, and Cognition,* **17,** 837–54.

Chapman, G. B. & Robbins, S. J. (1990). Cue interaction in human contingency judgment. *Memory and Cognition,* **18,** 537–45.

Chase, W. G. & Ericsson, K. A. (1981). Skilled memory. In J. R. Anderson (Ed.), *Cognitive skills and their acquisition* (pp. 141–89). Hillsdale, NJ:Erlbaum.

Chase, W. G. & Simon, H. A. (1973). Perception in chess. *Cognitive Psychology,* **4,** 55–81.

Chatlosh, D. L., Neunaber, D. J. & Wasserman, E. A. (1985). Response-outcome contingency: behavioral and judgmental effects of appetitive and aversive outcomes with college students. *Learning and Motivation,* **16,** 1–34.

Cheng, P. W. & Holyoak, K. J. (1995). Adaptive systems as intuitive statisticians: causality, contingency, and prediction. In J.-A. Meyer & H. Roitblat (Eds.), *Comparative approaches to cognition.* Cambridge, MA: MIT Press.

Cheng, P. W. & Novick, L. R. (1990). A probabilistic contrast model of causal induction. *Journal of Personality and Social Psychology,* **58,** 545–67.

Cheng, P. W. & Novick, L. R. (1992). Covariation in natural causal induction. *Psychological Review,* **99,** 365–82.

Davey, G. C. L. (1992a). An expectancy model of laboratory preparedness effects. *Journal of Experimental Psychology: General,* **121,** 24–40.

Davey, G. C. L. (1992b). Classical conditioning and the acquisition of human fears and phobias: a review and synthesis of the literature. *Advances in Behaviour Research and Therapy,* **14,** 29–66.

Davey, G. C. L. (1995). Preparedness and phobias: Specific evolved associations or a generalized expectancy bias? *Behavioral and Brain Sciences.* (In press.)

Deese, J. & Hulse, S. H. (1967). *The psychology of learning.* New York: McGraw-Hill.

DeLosh, E. L. (1993). *Interpolation and extrapolation in a functional learning paradigm* (Technical report). West Lafayette, IN: Purdue University, Purdue Mathematical Psychology Program.

Dickinson, A. (1980). *Contemporary animal learning theory.* Cambridge: Cambridge University Press.

Dickinson, A. (1985). Actions and habits: the development of behavioural autonomy. *Philosophical Transactions of the Royal Society of London,* **B308,** 67–78.

Dienes, Z. (1992). Connectionist and memory-array models of artificial grammar learning. *Cognitive Science,* **16,** 41–79.

Dienes, Z., Broadbent, D. E., & Berry, D. (1991). Implicit and explicit knowledge bases in artificial grammar learning. *Journal of Experimental Psychology: Learning, Memory, and Cognition,* **17,** 875–87.

Domjan, M. & Wilson, N. E. (1972). Specificity of cue to consequence in aversion learning in the rat. *Psychonomic Science,* **26,** 143–5.

Duda, R. O. & Hart, P. E. (1973). *Pattern classification and scene analysis.* New York: Wiley.

Dunn, J. C. & Kirsner, K. (1989). Implicit memory: task or process? In S. Lewandowsky, J. C. Dunn & K. Kirsner (Eds.), *Implicit memory: theoretical issues* (pp. 17–31). Hillsdale, NJ: Erlbaum.

Durlach, P. J. (1983). Effect of signaling intertrial unconditioned stimuli in autoshaping. *Journal of Experimental Psychology: Animal Behaviour Processes,* **9,** 374–9.

Ebbinghaus, H. (1885). *Memory: a contribution to experimental psychology.* New York: Dover.

Elman, J. L. (1990). Representation and structure in connectionist models. In G. T. M. Altmann (Ed.), *Cognitive models of speech processing: Psycholinguistic and computational perspectives* (pp. 345–82). Cambridge, MA: MIT Press.

Fales, E. & Wasserman, E. A. (1992). Causal knowledge: what can psychology teach philosophers? *Journal of Mind and Behavior*, **13**, 1–28.

Farley, J. (1987). Contingency learning and causal detection in *Hermissenda*: I. Behavior. *Behavioral Neuroscience*, **101**, 13–27.

Ghoneim, M. M., Block, R. I. & Fowles, D. C. (1992). No evidence of classical conditioning of electrodermal responses during anesthesia. *Anesthesiology*, **76**, 682–8.

Gibson, E. J. & Walk, R. D. (1956). The effect of prolonged exposure to visually presented patterns on learning to discriminate them. *Journal of Comparative and Physiological Psychology*, **49**, 239–42.

Gillund, G. & Shiffrin, R. M. (1984). A retrieval model for both recognition and recall. *Psychological Review*, **91**, 1–67.

Gluck, M. A. (1991). Stimulus generalization and representation in adaptive network models of category learning. *Psychological Science*, **2**, 50–5.

Gluck, M. A. & Bower, G. H. (1988). From conditioning to category learning: an adaptive network model. *Journal of Experimental Psychology: General*, **117**, 227–47.

Gluck, M. A. & Bower, G. H. (1990). Component and pattern information in adaptive networks. *Journal of Experimental Psychology: General*, **119**, 105–9.

Hall, G. (1991). *Perceptual and associative learning*. Oxford: Clarendon Press.

Hardin, C. L. (1990). Color and illusion. In W. G. Lycan (Ed.), *Mind and cognition* (pp. 555–67). Oxford: Blackwell.

Harlow, H. F. (1949). The formation of learning sets. *Psychological Review*, **56**, 51–65.

Hartley, J. & Homa, D. (1981). Abstraction of stylistic concepts. *Journal of Experimental Psychology: Human Learning and Memory*, **7**, 33–46.

Hays, W. L. (1963). *Statistics for psychologists*. New York: Holt, Rinehart and Winston.

Herrnstein, R. J. (1990). Levels of stimulus control: a functional approach. *Cognition*, **37**, 133–66.

Herrnstein, R. J., Vaughan, W., Mumford, D. B. & Kosslyn, S. M. (1989). Teaching pigeons an abstract relational rule: insideness. *Perception and Psychophysics*, **46**, 56–64.

Hintzman, D. L. (1976). Repetition and memory. In G. H. Bower (Ed.), *The psychology of learning and motivation* (Vol. **10**, pp. 47–91). New York: Academic Press.

Hintzman, D. L. & Ludlam, G. (1980). Differential forgetting of prototypes and old instances: simulation by an exemplar-based classification model. *Memory and Cognition*, **8**, 378–82.

Homa, D., Dunbar, S. & Nohre, L. (1991). Instance frequency, categorization, and the modulating effect of experience. *Journal of Experimental Psychology: Learning, Memory, and Cognition*, **17**, 444–458.

Homa, D., Sterling, S. & Trepel, L. (1981). Limitations of exemplar-based generalization and the abstraction of categorical information. *Journal of Experimental Psychology: Human Learning and Memory*, **7**, 418–39.

Hornik, K., Stinchcombe, M. & White, H. (1989). Multilayer feedforward networks are universal approximators. *Neural Networks*, **2**, 359–68.

Howard, R. W. (1993). On what intelligence is. *British Journal of Psychology*, **84**, 27–37.

Hunt, E. B., Marin, J. & Stone, P. J. (1966). *Experiments in induction*. San Diego: Academic Press.

Jacoby, L. L. (1991). A process dissociation framework: separating automatic from intentional uses of memory. *Journal of Memory and Language*, **30**, 513–41.

Jacoby, L. L. & Brooks, L. R. (1984). Nonanalytic cognition: memory, perception, and concept learning. In G. H. Bower (Ed.), *The psychology of learning and motivation* (Vol. **18**, pp. 1–47). New York: Academic Press.

James, W. (1890). *Principles of psychology*. New York: Holt.

Jared, D., McRae, K. & Seidenberg, M. S. (1990). The basis of consistency effects in word naming. *Journal of Memory and Language,* **29,** 687–715.

Jenkins, H. M. & Ward, W. C. (1965). Judgment of contingency between responses and outcomes. *Psychological Monographs,* **79** (Whole Number 594).

Knapp, A. G. & Anderson, J. A. (1984). Theory of categorization based on distributed memory storage. *Journal of Experimental Psychology: Learning, Memory, and Cognition,* **10,** 616–37.

Koffka, K. (1935). *Principles of Gestalt psychology.* New York: Harcourt Brace Jovanovic.

Koh, K. & Meyer, D. E. (1991). Function learning: Induction of continuous stimulus-response relations. *Journal of Experimental Psychology: Learning, Memory, and Cognition,* **17,** 811–36.

Kohler, W. (1947). *Gestalt psychology.* New York: Liveright.

Kohonen, T. (1977). *Associative memory: a system theoretical approach.* New York: Springer-Verlag.

Kripke, S. A. (1982). *Wittgenstein on rules and private language.* Oxford: Blackwell.

Kruschke, J. K. (1992). ALCOVE: an exemplar-based connectionist model of category learning. *Psychological Review,* **99,** 22–44.

Kruschke, J. K. (1993). Human category learning: implications for backpropagation models. *Connection Science,* **5,** 3–36.

Kushner, M., Cleeremans, A. & Reber, A. S. (1991). Implicit detection of event interdependencies and a PDP model of the process. In *Proceedings of the 13th annual conference of the Cognitive Science Society* (pp. 215–20). Hillsdale, NJ: Erlbaum.

Lipp, O. V., Siddle, D. A. T. & Vaitl, D. (1992). Latent inhibition in humans: single-cue conditioning revisited. *Journal of Experimental Psychology: Animal Behavior Processes,* **18,** 115–25.

Lovibond, P. F. (1992). Tonic and phasic electrodermal measures of human aversive conditioning with long duration stimuli. *Psychophysiology,* **29,** 621–32.

Lubow, R. E. (1989). *Latent inhibition and conditioned attention theory.* Cambridge: Cambridge University Press.

Luria, A. R. (1968). *The mind of a mnemonist.* Cambridge, MA: Harvard University Press.

McClelland, J. L. & Rumelhart, D. E. (1985). Distributed memory and the representation of general and specific information. *Journal of Experimental Psychology: General,* **114,** 159–88.

McClelland, J. L. & Rumelhart, D. E. (1988). *Explorations in parallel distributed processing.* Cambridge, MA: MIT Press.

McCloskey, M. & Cohen, N. J. (1989). Catastrophic interference in connectionist networks: the sequential learning problem. In G. H. Bower (Ed.), *The psychology of learning and motivation* (Vol. **24,** pp. 109–65). San Diego: Academic Press.

MacFarlane, D. A. (1930). The role of kinesthesis in maze learning. *University of California Publications in Psychology,* **4,** 277–305.

Mackintosh, N. J. (1988). Approaches to the study of animal intelligence. *British Journal of Psychology,* **79,** 509–25.

McLaren, I. P. L., Kaye, H. & Mackintosh, N. J. (1989). An associative theory of the representation of stimuli: Applications to perceptual learning and latent inhibition. In R. G. M. Morris (Ed.), *Parallel distributed processing: Implications for psychology and neurobiology* (pp. 102–30). Oxford: Oxford University Press.

McLaren, I. P. L., Leevers, H. J. & Mackintosh, N. J. (1994). Recognition, categorization, and perceptual learning (or, How learning to classify things together helps one to tell them apart). In C. Umilta & M. Moscovitch (Eds.), *Attention and Performance XV: Conscious and nonconscious information processing* (pp. 889–909). Cambridge, MA: MIT Press.

McNally, R. J. (1987). Preparedness and phobias. *Psychological Bulletin,* **101,** 283–303.

Malt, B. C. (1989). An on-line investigation of prototype and exemplar strategies in classification. *Journal of Experimental Psychology: Learning, Memory, and Cognition,* **15,** 539–55.

Marr, D. (1982). *Vision: A computational investigation into the human representation and processing of visual information*. San Francisco: Freeman.

Mathews, R. C., Buss, R. R., Stanley, W. B., Blanchard-Fields, F., Cho, J. R. & Druhan, B. (1989). Role of implicit and explicit processes in learning from examples: a synergistic effect. *Journal of Experimental Psychology: Learning, Memory, and Cognition*, **15**, 1083–100.

Medin, D. L. (1975). A theory of context in discrimination learning. In G. H. Bower (Ed.), *The psychology of learning and motivation* (Vol. 9, pp. 263–314). New York: Academic Press.

Medin, D. L., Altom, M. W., Edelson, S. M. & Freko, D. (1982). Correlated symptoms and simulated medical classification. *Journal of Experimental Psychology: Learning, Memory, and Cognition*, **8**, 37–50.

Medin, D. L. & Edelson, S. M. (1988). Problem structure and the use of base-rate information from experience. *Journal of Experimental Psychology: General*, **117**, 68–85.

Medin, D. L. & Schaffer, M. M. (1978). Context theory of classification learning. *Psychological Review*, **85**, 207–38.

Medin, D. L. & Schwanenflugel, P. J. (1981). Linear separability in classification learning. *Journal of Experimental Psychology: Human Learning and Memory*, **7**, 355–68.

Medin, D. L. & Smith, E. E. (1981). Strategies and classification learning. *Journal of Experimental Psychology: Human Learning and Memory*, **7**, 241–53.

Melz, E. R., Cheng, P. W., Holyoak, K. J. & Waldmann, M. R. (1993). Cue competition in human categorization: contingency or the Rescorla-Wagner learning rule? Comment on Shanks (1991). *Journal of Experimental Psychology: Learning, Memory, and Cognition*, **19**, 1398–410.

Metcalfe, J. & Fisher, R. P. (1986). The relation between recognition memory and classification learning. *Memory and Cognition*, **14**, 164–73.

Minsky, M. L. & Papert, S. A. (1969). *Perceptrons: An introduction to computational geometry*. Cambridge, MA: MIT Press.

Morgan, J. L., Meier, R. P. & Newport, E. L. (1989). Facilitating the acquisition of syntax with cross-sentential cues to phrase structure. *Journal of Memory and Language*, **28**, 360–74.

Nakamura, G. V. (1985). Knowledge-based classification of ill-defined categories. *Memory and Cognition*, **13**, 377–84.

Neunaber, D. J. & Wasserman, E. A. (1986). The effect of unidirectional versus bidirectional rating procedures on college students' judgments of response-outcome contingency. *Learning and Motivation*, **17**, 162–79.

Newell, A., Shaw, J. C., & Simon, H. A. (1958). Elements of a theory of human problem solving. *Psychological Review*, **65**, 151–66.

Nosofsky, R. M. (1986). Attention, similarity and the identification-categorization relationship. *Journal of Experimental Psychology: General*, **115**, 39–57.

Nosofsky, R. M. (1987). Attention and learning processes in the identification and categorization of integral stimuli. *Journal of Experimental Psychology: Learning, Memory, and Cognition*, **13**, 87–108.

Nosofsky, R. M. (1988). Exemplar-based accounts of relations between classification, recognition, and typicality. *Journal of Experimental Psychology: Learning, Memory, and Cognition*, **14**, 700–8.

Nosofsky, R. M. (1992). Similarity scaling and cognitive process models. *Annual Review of Psychology*, **43**, 25–53.

Nosofsky, R. M., Clark, S. E. & Shin, H. J. (1989). Rules and exemplars in categorization, identification, and recognition. *Journal of Experimental Psychology: Learning, Memory, and Cognition*, **15**, 282–304.

Nosofsky, R. M., Gluck, M. A., Palmeri, T. J., McKinley, S. C. & Glauthier, P. (1994). Comparing models of rule-based classification learning: a replication and extension of Shepard, Hovland, and Jenkins (1961). *Memory and Cognition*, **22**, 352–69.

Nosofsky, R. M. & Kruschke, J. K. (1992). Investigations of an exemplar-based connectionist model of category learning. In D. L. Medin (Ed.), *The psychology of learning and motivation* (Vol. **28**, pp. 207–50).

Nosofsky, R. M., Kruschke, J. K. & McKinley, S. C. (1992). Combining exemplar-based category representations and connectionist learning rules. *Journal of Experimental Psychology: Learning, Memory, and Cognition,* **18**, 211–33.

Perruchet, P. (1994). Learning from complex rule-governed environments: on the proper functions of nonconscious and conscious processes. In C. Umilta & M. Moscovitch (Eds.), *Attention and Performance XV: Conscious and nonconscious information processing* (pp. 811–35). Cambridge, MA: MIT Press.

Peterson, C. R. & Beach, L. R. (1967). Man as an intuitive statistician. *Psychological Bulletin,* **68**, 29–46.

Pinker, S. (1991). Rules of language. *Science,* **253**, 530–5.

Plunkett, K. & Marchman, V. (1991). U-shaped learning and frequency effects in a multi-layered perceptron: Implications for child language acquisition. *Cognition,* **38**, 43–102.

Posner, M. I. (1964). Information reduction in the analysis of sequential tasks. *Psychological Review,* **71**, 491–504.

Postman, L. (1962). Repetition and paired-associate learning. *American Journal of Psychology,* **75**, 372–89.

Quillian, M. R. (1968). Semantic memory. In M. Minsky (Ed.), *Semantic information processing.* Cambridge, MA: MIT Press.

Reber, A. S. (1967). Implicit learning of artificial grammars. *Journal of Verbal Learning and Verbal Behavior,* **5**, 855–63.

Reber, A. S. (1989). Implicit learning and tacit knowledge. *Journal of Experimental Psychology: General,* **118**, 219–35.

Reed, P. (1992). Effect of local context of responding on human judgment of causality. *Memory and Cognition,* **20**, 573–9.

Reed, P. (1993). Influence of the schedule of outcome presentation on causality judgments. *Quarterly Journal of Experimental Psychology,* **46A**, 327–45.

Regehr, G. & Brooks, L. R. (1993). Perceptual manifestations of an analytic structure: the priority of holistic individuation. *Journal of Experimental Psychology: General,* **122**, 92–114.

Reiss, S. & Wagner, A. R. (1972). CS habituation produces a 'latent inhibition effect' but no active conditioned inhibition. *Learning and Motivation,* **3**, 237–45.

Rescorla, R. A. (1968). Probability of shock in the presence and absence of CS in fear conditioning. *Journal of Comparative and Physiological Psychology,* **66**, 1–5.

Rescorla, R. A. & Wagner. A. R. (1972). A theory of Pavlovian conditioning: Variations in the effectiveness of reinforcement and nonreinforcement. In A. H. Black & W. F. Prokasy (Eds.), *Classical conditioning II: current theory and research* (pp. 64–99). New York: Appleton-Century-Crofts.

Restle, F. (1965). Significance of all-or-none learning. *Psychological Bulletin,* **64**, 313–25.

Roediger, H. L. (1993). Learning and memory: progress and challenge. In D. E. Meyer & S. Kornblum (Eds.), *Attention & Performance XIV: Synergies in experimental psychology, artificial intelligence, and cognitive neuroscience* (pp. 509–28). Cambridge, MA: MIT Press.

Romanes, G. J. (1882). *Animal intelligence.* London: Kegan, Paul, Trench & Co.

Rosch, E. H. (1973). On the internal structure of perceptual and semantic categories. In T. E. Moore (Ed.), *Cognitive development and the acquisition of language* (pp. 111–44). New York: Academic Press.

Rosch, E., Simpson, C. & Miller, R. S. (1976). Structural bases of typicality effects. *Journal of Experimental Psychology: Human Learning and Memory,* **2**, 491–502.

Rosenblatt, F. (1962). *Principles of neurodynamics.* New York: Spartan.

Sacks, O. (1992). The last hippie. *New York Review,* March 26, pp. 51–60.

Salmon, W. C. (1984). *Scientific explanation and the causal structure of the world.* Princeton, NJ: Princeton University Press.

Schacter, D. L., Eich, J. E. & Tulving, E. (1978). Richard Semon's theory of memory. *Journal of Verbal Learning and Verbal Behavior*, **17**, 721–43.

Seligman, M. E. P. (1971). Phobias and preparedness. *Behavior Therapy*, **2**, 307–20.

Shallice, T. (1988). *From neuropsychology to mental structure.* Cambridge: Cambridge University Press.

Shanks, D. R. (1985). Forward and backward blocking in human contingency judgement. *Quarterly Journal of Experimental Psychology*, **37B**, 1–21.

Shanks, D. R. (1986). Selective attribution and the judgment of causality. *Learning and Motivation*, **17**, 311–34.

Shanks, D. R. (1989). Selectional processes in causality judgment. *Memory and Cognition*, **17**, 27–34.

Shanks, D. R. (1990). Connectionism and the learning of probabilistic concepts. *Quarterly Journal of Experimental Psychology*, **42A**, 209–37.

Shanks, D. R. (1991a). Categorization by a connectionist network. *Journal of Experimental Psychology: Learning, Memory, and Cognition*, **17**, 433–43.

Shanks, D. R. (1991b). Some parallels between associative learning and object classification. In J.-A. Meyer & S. W. Wilson (Eds.), *From animals to animats: proceedings of the first international conference on simulation of adaptive behavior* (pp. 337–43). Cambridge, MA: MIT Press.

Shanks, D. R. (1992). Connectionist accounts of the inverse base-rate effect in categorization. *Connection Science*, **4**, 3–18.

Shanks, D. R. & Dickinson, A. (1991). Instrumental judgment and performance under variations in action-outcome contingency and contiguity. *Memory and Cognition*, **19**, 353–60.

Shanks, D. R. & St. John, M. F. (1994). Characteristics of dissociable human learning systems. *Behavioral and Brain Sciences*, **17**, 367–447.

Shanks, D. R., Pearson, S. M. & Dickinson, A. (1989). Temporal contiguity and the judgment of causality. *Quarterly Journal of Experimental Psychology*, **41B**, 139–59.

Shepard, R. N. (1958). Stimulus and response generalization: tests of a model relating generalization to distance in psychological space. *Journal of Experimental Psychology*, **55**, 509–23.

Shepard, R. N. (1980). Multidimensional scaling, tree-fitting, and clustering. *Science*, **210**, 390–8.

Shepard, R. N. (1987). Toward a universal law of generalization for psychological science. *Science*, **237**, 1317–1323.

Shepard, R. N., Hovland, C. L. & Jenkins, H. M. (1961). Learning and memorization of classifications. *Psychological Monographs*, **75**, whole no. 517.

Sidman, M. & Tailby, W. (1982). Conditional discrimination vs. matching to sample: an expansion of the testing paradigm. *Journal of the Experimental Analysis of Behavior*, **37**, 5–22.

Smedslund, J. (1963). The concept of correlation in adults. *Scandinavian Journal of Psychology*, **4**, 165–73.

Smith, E. E., Langston, C. & Nisbett, R. E. (1992). The case for rules in reasoning. *Cognitive Science*, **16**, 1–40.

Smith, L. B. (1989). From global similarities to kinds of similarities: the construction of dimensions in development. In S. Vosniadou & A. Ortony (Eds.), *Similarity and analogical reasoning* (pp. 146–78). New York: Cambridge University Press.

Solomon, R. L. & Turner, L. H. (1962). Discriminative classical conditioning in dogs paralyzed by curare can later control discriminative avoidance responses in the normal state. *Psychological Review*, **69**, 202–19.

Sutton, R. S. & Barto, A. G. (1981). Toward a modern theory of adaptive networks: Expectation and prediction. *Psychological Review*, **88**, 135–70.

Tomarken, A. J., Mineka, S. & Cook, M. (1989). Fear-relevant selective associations and covariation bias. *Journal of Abnormal Psychology*, **98**, 381–94.

Tulving, E. (1972). Episodic and semantic memory. In E. Tulving & W. Donaldson (Eds.), *Organization and memory* (pp. 381–403). New York: Academic Press.

Tulving, E. (1983). *Elements of episodic memory.* Oxford: Oxford University Press.

Underwood, B. J. (1957). Interference and forgetting. *Psychological Review, 64,* 49–60.

Valentine, T. & Bruce, V. (1986). The effects of distinctiveness in recognising and classifying faces. *Perception,* 15, 525–35.

Van Hamme, L. J. & Wasserman, E. A. (1994). Cue competition in causality judgments: the role of nonpresentation of compound stimulus elements. *Learning and Motivation,* 25, 127–51.

Vaughan, W. (1988). Formation of equivalence sets in pigeons. *Journal of Experimental Psychology: Animal Behavior Processes,* 14, 36–42.

Vokey, J. R. & Brooks, L. R. (1992). Salience of item knowledge in learning artificial grammars. *Journal of Experimental Psychology: Learning, Memory, and Cognition,* 18, 328–44.

Waldmann, M. R. & Holyoak, K. J. (1992). Predictive and diagnostic learning within causal models: Asymmetries in cue competition. *Journal of Experimental Psychology: General,* 121, 222–36.

Ward, J. (1893). Assimilation and association. *Mind,* 2, 347–62.

Wasserman, E. A., Chatlosh, D. L. & Neunaber, D. J. (1983). Perception of causal relations in humans: factors affecting judgments of response-outcome contingencies under free-operant procedures. *Learning and Motivation,* 14, 406–32.

Wasserman, E. A., Elek, S. M., Chatlosh, D. L. & Baker, A. G. (1993). Rating causal relations: the role of probability in judgments of response-outcome contingency. *Journal of Experimental Psychology: Learning, Memory, and Cognition,* 19, 174–88.

Watkins, M. J. & Kerkar, S. P. (1985). Recall of a twice-presented item without recall of either presentation: generic memory for events. *Journal of Memory and Language,* 24, 666–78.

Whittlesea, B. W. A. (1987). Preservation of specific experiences in the representation of general knowledge. *Journal of Experimental Psychology: Learning, Memory, and Cognition,* 13, 3–17.

Whittlesea, B. W. A. & Dorken, M. D. (1993). Incidentally, things in general are particularly determined: an episodic-processing account of implicit learning. *Journal of Experimental Psychology: General,* 122, 227–48.

Widrow, B. & Hoff, M. E. (1960). Adaptive switching circuits. *1960 IRE WESCON Convention Record (Pt. 4),* pp. 96–104.

Williams, D. A. (1995). Forms of inhibition in animal and human learning. *Journal of Experimental Psychology: Animal Behavior Processes.* 21, 129–142.

Williams, D. A., Sagness, K. E. & McPhee, J. E. (1994). Configural and elemental strategies in predictive learning. *Journal of Experimental Psychology: Learning, Memory, and Cognition,* 20, 694–709.

Willingham, D. B., Greeley, T. & Bardone, A. M. (1993). Dissociation in a serial response time task using a recognition measure: comment on Perruchet and Amorim (1992). *Journal of Experimental Psychology: Learning, Memory, and Cognition,* 19, 1424–30.

Willshaw, D. J., Buneman, O. P. & Longuet-Higgins, H. C. (1969). Non-holographic associative memory. *Nature,* 222, 960–2.

Wilson, P. (1993). *Categorisation and configural learning.* Paper presented at the Experimental Analysis of Behaviour Group conference, London, 29–31 March.

Wittgenstein, L. (1958). *Philosophical investigations.* Oxford: Blackwell.

Wixted, J. T. & Ebbesen, E. B. (1991). On the form of forgetting. *Psychological Science,* 2, 409–15.

Young, R. M. & O'Shea, T. (1981). Errors in children's subtraction. *Cognitive Science,* 5, 153–77.

Zimmer-Hart, C. L. & Rescorla, R. A. (1974). Extinction of Pavlovian conditioned inhibition. *Journal of Compararive and Physiological Psychology,* 86, 837–45.

Index

Abramson, L. Y., 24
acquisition curves, 31–3, 110–12, 114–17
action–outcome learning, 13, 20, 147
Allan, L. G., 22
Allen, S. W., 13, 26–33, 88, 158
Alloy, L. B., 23–4
Altmann, G. T. M., 168–70
Altom, M. W., 73–4
amnesia, 1, 14, 100
anaesthesia, 17
Anderson, J. A., 105, 121
Anderson, J. R., 10, 12, 50
animal
  conditioning, 25, 34, 107, 140
  learning, 4, 58, 98, 105
artificial grammar learning, 15–16, 18,
    88–91, 108–9, 125–6, 144, 167
Ashby, F. G., 74–6
associationists, 5–10, 12, 19, 33, 53, 96, 102,
    104–5
associative learning, 2, 5
asymptote of learning, 31–3, 37–8, 48, 59,
    112, 114–15, 118, 120
attention, 138, 145
Attneave, F., 144
Austin, G. A., 8

Baker, A. G., 30–1, 37, 43, 48–9, 55,
    114–15
Bardone, A. M., 14, 17
Barsalou, L. W., 9, 81
Barto, A. G., 107
base-rate, 97, 99, 121, 124
Bayes theorem, 33
Beach, L. R., 26
behaviourists, 3
Berry, D., 16, 18–19
bias, 24, 26, 32, 37, 50, 52–4, 59–60, 124–5,
    143
  response, 66
Bitterman, M. E., 7
Blanchard-Fields, F., 15
Block, R. I., 17
blocking, 47, 118, 142–3

backward, 147–9
  forward, 148
Bower, G. H., 96, 99, 121, 124, 149
Broadbent, D. E., 16, 18–19, 150–1
Brooks, L. R., 13, 76–7, 89–90, 126,
    158–61, 164, 168
Bruce, V., 70–1
Bruner, J. S., 8
Buneman, O. P., 105
Buss, R. R., 15

categorisation, 68, 70, 78, 86–8, 102
category learning, 13, 81–2, 103
causal selection, 48
Chandler, C. C., 93–5
Chapman, G. B., 46–8, 52–4, 112, 114,
    117–19, 141–2, 146–50
Chase, W. G., 104
Chatlosh, D. L., 25, 29–31, 37, 39, 43, 55,
    114–15, 117–18
Cheng, P. W., 10, 20, 40, 43, 45, 48–50, 52,
    59, 118
Cho, J. R., 15
choice responses, 13
Clark, S. E., 161–5, 167, 178
classification, 68–78, 81–90, 92, 99, 102,
    108, 124, 127, 129, 135, 137–8, 140,
    153, 161, 171, 177
  filtration/condensation effect, 137–8, 140
  latencies, 160
  linearly separable, 131
  nonlinearly separable, 127, 131, 139, 149,
    151
  rule based, 158, 160
  similarity, 134,
Cleeremans, A., 155–6
cognitive theory of learning, 3
Cohen, N. J., 132–4
concepts, 8
  abstract, 179–80
  learning, 81, 102, 121
  relational, 176, 180
  representation of, 8
conditional discrimination procedure, 174